Western Stock Ranching

WESTERN STOCK RANCHING

by

Mont H. Saunderson

RANGE ECONOMIST, U.S. FOREST SERVICE
FORMERLY RANCH ECONOMIST, MONTANA STATE COLLEGE

THE UNIVERSITY OF MINNESOTA PRESS, Minneapolis
LONDON · GEOFFREY CUMBERLEGE · OXFORD UNIVERSITY PRESS

First Edition

TO ALL WHO STRIVE TO MANAGE THE RANGELANDS OF
THE WEST FOR A SUSTAINED PRODUCTIVE ECONOMY
THIS BOOK IS DEDICATED

Preface

In the following presentation of materials concerning western stock ranch economy and management, my purpose is to show the practical applications of these materials to the planning and management of the stock ranch.

In presenting this set of materials covering the stock ranch economy of all the western states, it is of course realized that wide differences prevail in the way ranches must operate. To a large degree, however, these differences can be resolved in terms of the several major natural regions of the West. Within these regions the natural and economic factors influencing ranch management can be satisfactorily depicted and the main types of ranches and their management standards satisfactorily analyzed. The first chapter develops this regional characterization, which forms a background for all the succeeding chapters.

In several of these chapters considerable attention is given to land resource economics — the changes in the resource and how such changes affect the economy of the ranches. Any competent observer is impressed with the extensive change on western rangelands from the perennial range plants to the annuals and with the consequent effects upon the economy of the ranches. Rather generally in the past the management programs of western stock ranches have given primary emphasis to livestock husbandry and secondary rather than coordinate attention to the problems of the land resource. It is apparent that for some types of western rangeland resources the livestock production can be well maintained for a considerable period of time before the economic consequences of resource deterioration become clearly apparent. As a consequence the livestock production has not been a good guide for judging the condition and trend of the resource.

In Chapters II, III, and IV emphasis has been placed upon the interrelationships of livestock management, rangeland management, and the use of feed crops in the production plan of the ranch as an operating unit. Chapter V shows the relationship of markets and prices to the production planning and gives the production and income standards for different regions and types of ranches. Chapter VI deals primarily with production cost standards, methods of cost analysis, and financial planning. Chapter VII concerns the long-time management plan of the ranch and the relationship of such a plan to the annual production planning. The final chapter shows some of the relationships of federal public land use and federal public land policy to the management of stock ranches that use such lands.

Many of the materials used throughout these chapters have been developed through field surveys and ranch management studies involving close working contacts with ranches and with ranch operators. Some of the cost data were obtained from the files of production credit associations. This publication is prepared primarily for applied use by ranch operators, workers in finance, science, education, and administration, and by others interested in the permanence and success of western stock ranches.

I wish to acknowledge valuable aid from several of my associates in the U.S. Forest Service, especially W. L. Dutton, E. P. Cliff, G. D. Pickford, Glen A. Smith (now retired), and P. V. Woodhead (deceased). Forrest Bassford of the *Western Livestock Journal* and B. W. Allred of the U.S. Soil Conservation Service helped me with certain of the materials. R. B. Tootell and others of my former associates at Montana State College are among the many to whom I am indebted for help, advice, and encouragement.

The responsibility for the materials and statements presented in this book rests entirely with me and nothing contained herein should in any way be construed as an expression of the official viewpoint of the government service with which I am connected.

Mont H. Saunderson

Denver, Colorado
July 1, 1950

Contents

ix

LIST OF TABLES

LIST OF FIGURES

LIST OF FORMS

CONTENTS

Western Stock Ranching

CHAPTER I

Resources and Ranches

"Western stock ranch operations depend a lot on how the country is put together."

This salty observation of a western ranchman suggests the sweeping vastness and diversity of nature "west of the hundredth" and implies the many resulting contrasts in western ranches and ranching.

But across the immense lands of the western plains, mountains, valleys, and deserts lies a clear design of natural regions, a vast but distinct pattern of the lands and of the ranches that use the lands. This pattern is a fascinating one for those who enjoy the western out-of-doors; comprehension of it can be useful to anyone who seeks a better understanding of western rangelands and ranching.

Let's take a look at this pattern. You have a map of the natural regions in Figure 1.

We'll first turn our view to the Great Plains. This great expanse of grasslands reaches from Canada to Mexico. The hundredth meridian and the Rocky Mountain chain form the approximate eastern and western boundaries. One of the world's most productive natural grasslands, the Great Plains regions account for about half the rangeland grazing capacity of the West. These plains are often called the "short-grass country," but that is really an oversimplification. Great Plains rangelands are best comprehended in terms of three natural regions.

Northern Plains Region. In the northern plains — the Platte River marks the approximate southern boundary — we see the "mixed-prairie" type of range, a mixture of the midgrasses and

3

the short grasses. The midgrasses, ancestrally cool-climate grasses of intermediate size, grow early in the season and again make some new growth in the autumn. Northern plains midgrasses are principally the wheat and the needle grasses. Balancing this spring and fall growth of the midgrasses, the short grasses (warm-climate grasses) grow rapidly in the heat of summer when topsoil mois-

FIGURE 1. RANGE RESOURCE AND RANCH MANAGEMENT REGIONS OF THE WESTERN UNITED STATES

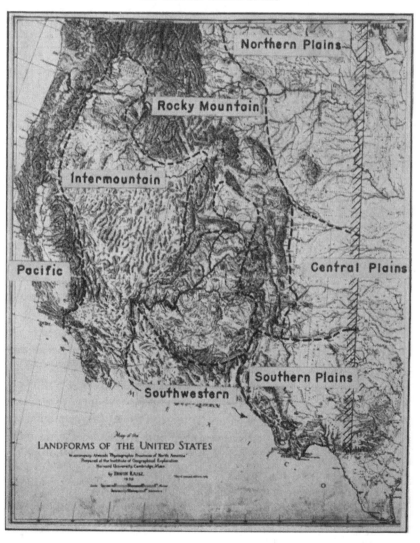

ture is available. Most important of the northern plains short grasses are the blue grama and the buffalograss.

When in good condition the mixed-prairie type of rangeland provides well-balanced grazing through the nine- or ten-month range season and affords native hay for winter feed. The average grazing capacity of this range is about three acres per animal month for cattle, three fourths of an acre per animal month for sheep. Overuse of this range diminishes the midgrasses more than the short grasses, turning it into a short-grass type. The consequence? An unbalancing of the seasonal capacities and a diminished total capacity. Since the short grasses are good sod-formers, this range endures heavy use, but continued heavy use doesn't pay.

Though naturally suited to cattle somewhat better than to sheep, northern plains rangelands support a large population of both kinds of livestock. Cattle predominate in the northern plains roughlands between the Yellowstone and the North Platte. Here the prevalence of midgrasses and the natural shelter afforded by rough terrain and by trees and brush along the streams definitely favor range cattle production. Cattle ranches in these roughlands usually have to do only a limited amount of winter feeding, thirty to ninety days of intermittent feeding through storm periods. Native hay, grain hay, and dry-land corn provide the winter feed. Feeding requirements have increased some with the shift of these ranches toward a breeding herd basis and the marketing of younger animals to the Corn Belt feeders.

North of the Missouri River lie the flat glacial plains, typified by the plains of the Milk River in northern Montana. Here the northern plains do not as a rule have adequate natural shelter. Extensive range sheep operations use the Milk River plains through the range season and then move considerable distances to the river valley for sheltered winter feeding grounds and for the alfalfa hay grown under irrigation in the valley. These sheep operations use the fine-wool breeds of sheep, lamb late in the spring after fresh range growth has started, and market feeder lambs weighing about sixty-five pounds when sold in October. Mutton-breed sheep and range-fat lamb production takes more green feed than these ranges provide in late summer.

Large and spectacular is that biological island of the northern

Naturally well-balanced midgrasses and short grasses make the mixed-prairie range type of the northern plains. (U.S. Soil Conservation Service Photo)

Nature is a bit touchy in the Nebraska sandhills, but bountiful when well treated. (U.S. Forest Service Photo)

6

plains known as the Nebraska sandhills. Here we see some 15 million acres of rangeland "haired out" with the tall grasses. This island of tall grasses in the midgrass and short-grass range plant association of the northern plains lies between the Platte and Niobrara rivers in northwestern Nebraska. The permanent deep-soil moisture of the sandhills favors the tall grasses of the humid country. Some of the midgrasses are found also, but the usually dry topsoil of the sand just about excludes the shallow-rooted short grasses. Because of the dune-sand soils, nature is a bit touchy in the sandhills, but bountiful when well treated. Too intensive grazing of this range gives the winds a bite at the loose sand, and that means trouble!

A cowman's domain, the sandhills, as you have observed if you have driven the sand-trail roads ("paved" with native hay on the soft spots and the grades). Those coarse grasses, the turkeyfoot and the other bluestems, wave a "keep out" to the sheepman; the soft sandy soils give an equally effective warning to the farmer.

Great expanses of fenced ranges, herds of well-bred beef cattle, windmills, and an occasional ranch home dominate the landscape from North Platte to Valentine. Cherry County — area 6000 square miles, rural population 8000 — boasts of more cattle than the entire state of Wyoming. It is a pardonable overstatement, however. Cherry County's cattle have at times reached the quarter million figure; Wyoming's recent beef cattle population totaled about one million.

Recent surveys indicate that during the war period the sandhills ranches increased their cattle to the extent that range use averaged about one and a half acres per animal month — too heavy for sustained good results, say the range technicians of the U.S. Forest Service. The Forest Service has two small units of national forest land in the sandhills for afforestation and grazing observation and demonstration.

Central Plains Region. We turn southward now from the Platte River and see the transition from northern plains mixed-prairie ranges to the short-grass country of the central plains region. Sod-forming short grasses — buffalograss, blue grama, and some of the southwestern gramas — clothe the lands here. The central plains stretch continuously southward to northern Texas and northeastern New Mexico, where we see the southern limits of the high plains at the "break of the plains" or escarpment of the Ogallala. Far to the western horizon we see the central plains sweep to the front range of the Rocky Mountains, in contrast with the several mountain and roughland areas that break the expanses of the northern plains.

Limited or absent are the cool-climate grasses, the midgrasses we observed in the northern plains. But the short grasses, nearer in the central plains to their original southwestern climate, grow vigorously and in variety. This short-grass range provides a good dry feed after it cures in late summer. But after these short grasses cure, they are a fattening feed rather than the milk-producing feed needed by cows with calves. Since the trend for all western cattle ranches is toward breeding herds and feeder-calf marketing, central plains cattle ranches have adapted more to stock farming in making this change. Range feed is fully used through the summer, then balanced with cropland pasturages and feeds through autumn and winter. Fall-sown wheat, dry-land corn, grain sorghums, and Sudan grass constitute the crop feeds used to balance the short-grass summer grazing.

Another reason for the modification of central plains ranches to stock farms has been the extensive breaking of the native grass-

Central plains short-grass ranges contrast with the mixed-prairie type of the northern plains. (U.S. Soil Conservation Service Photo)

Southern plains ranges are a complex association of plains short grasses, desert grasses, shrubs, and early-growing annuals. (U.S. Soil Conservation Service Photo)

lands for wheat farming. Central plains short-grass ranges afford good sheep grazing, but the dry-land crops that can be grown for winter feeds are coarse roughages not too well adapted to sheep. We do not, consequently, see many sheep ranches. In the favorable seasons for fall and early winter pasturage on the fall-sown wheat the lamb feeders move large numbers of feeder lambs into the central plains from the west for fattening on the wheat pasturage. This fairly recent development has been quite successful. In those seasons favorable for this kind of pasturage a high percentage of the lambs fatten sufficiently for slaughter without any feedlot finishing.

Southern Plains Region. We turn our view now to the southern plains, the great ranching empire of central and western Texas and southeastern New Mexico. Again we see a transition in the rangelands, a new and different set of the natural influences so important to the ranches. Here we find the range plant life a complex association of the plains short grasses, the southwestern semidesert grasses such as curly mesquite and galleta, several leguminous shrubs, and a number of winter-growing annuals. This very productive range will, when in good condition, provide well-balanced, yearlong grazing for both cattle and sheep. Crop feeds are used only to a limited extent for winter maintenance of range animals in the southern plains, but feeding of a protein concentrate as a winter range supplement has increased markedly in recent years. Cottonseed meal fed on the range during the winter in cake or pellet form makes an effective addition to the range feed.

Because of his yearlong grazing the southern plains ranchman figures his land grazing capacity in number of animals per 640-acre section of land rather than in acres per animal month. Many of these rangelands will, when in good condition, carry 30 or more head of cattle or 150 sheep per section of rangeland. Some rangelands do better than this, others are considerably below these figures.

The number of sheep in this region was just about tripled between 1920 and 1940, while during this time the number of cattle was only slightly reduced. Thus, in the general picture these rangelands are carrying a heavy load, and the results are becoming apparent. Large areas of central and western Texas that were predominantly grasslands are shifting more to shrubs, low trees, cacti, and weeds. There may be some other influences in this not as yet too well understood (as illustrated by fire suppression since the occupation of these lands), but heavy grazing appears to be the main influence in this change.

Southern plains ranches carry some surprisingly high capital values, especially when they are well fenced and provided with adequate water facilities. These high ranch investments are both compensated by and due to the low requirements and costs for labor and crop feeds. Investments in land and improvements totaling $150 per head of ranch capacity for cattle and $30 for sheep were common before the recent war period. This land capitalization is approximately double what has been considered a good general standard elsewhere in the West.

Rather generally the southern plains ranches fence and cross-fence the rangelands in a system of range pastures or paddocks to facilitate land and livestock management. Where this system prevails sheep are not herded in bands, the usual practice of other western range sheep operations. Southern plains sheep ranches probably adhere to the fine-wool breeds more than the range sheep operations elsewhere in the West do. Many of the cattle ranches in the lower and hotter parts of the region cross Brahman cattle with American beef breeds to increase resistance to the parasites which plague livestock in warm climates.

Rocky Mountain Region. We turn now from the plains to the Rocky Mountain region, the region of mountain valley and foothill ranches. From the Canadian border this region runs through Idaho and western Montana and Wyoming, includes the Wasatch and Uintah mountains of northern Utah, extends through central Colorado and New Mexico, and ends in the vicinity of Santa Fe.

Though this region extends through more than 1000 miles of latitude, we see throughout its length the same natural influences for ranch operations. Here we have our introduction to natural "zones" of range plant life. As we move up the elevation scale from valley floors to the high country, we see how the changes in moisture, temperature, and winter snow cover result in several distinctly different natural vegetative zones or range types. These natural zones spell distinct differences in seasons of range use and in suitability for cattle or sheep. Let's follow the progression of these zones up the mountain, observing how seasonal ranges fit into the operating pattern of the ranches.

Lowest in elevation, a zone of grassland extends from the valley floor up the lower foothills to the point where the ponderosa pine (western yellow pine) growth begins. In this grassland zone we have our first look at the bunchgrasses, which, in contrast to the plains grasses, do not usually form a sod. This range type is well suited for cattle range, spring and fall or spring, summer, and fall. Where this range has a good complement of the smaller and finer bunchgrasses such as the Idaho fescue, it is suited to sheep. It is, however, used only as a spring and fall sheep range where the uplands afford the green summer feed for putting bloom and finish on the lambs. The average grazing capacity of the foothill grassland ranges, when they are in good condition, usually runs from two to three acres per animal month for cattle and one-half to one acre per animal month for sheep. These lands, along with the valley bottom lands, usually constitute the "deeded," or operation-owned, lands of the ranches. Most of the mountain lands from the start of the western yellow pine growth are national forest lands.

Above the grasslands of the lower foothills lies the ponderosa pine zone or "pine-bunchgrass" range type. This zone lies between the 4000- and 6000-foot elevations in the northern parts of the re-

Rocky Mountain foothill grasslands lie between the valley floors and the western yellow pine zone of the lower mountain slopes. (Glen A. Smith Photo)

Above the Rocky Mountain foothill grasslands lies the ponderosa pine zone or pine-bunchgrass range type. (U.S. Forest Service Photo)

12

gion and somewhat higher in the southern parts. Bunchgrass and browse plants in the timber stands and in the open spaces or "parks" constitute the range feed in this zone. Because of a prevalence of the coarser bunchgrasses in this zone the range feed is best suited to cattle. The elevation usually limits the season to the months between late May and early October.

As we move upland from the ponderosa pine zone, we see the spruce-fir zone extending upward to the 8000- or 9000-foot level. Spruce and fir trees give the general aspect to this zone. Range feed consists of weeds, grasses, sedges, and shrubs in the timber stands, parks, and mountain meadows along the streams. Generally better suited to sheep than to cattle, this range type provides summer grazing for both cattle and sheep, but usually the latter.

Above the spruce-fir zone and above the timber line lies the alpine country. Here we see alpine clovers, short sedges, small bunchgrasses, and weeds growing on the thin soil mantle that nature has developed in this land of short growing seasons. Range forage of the alpine country, suited to sheep rather than cattle, can be used only for a short summer season of two or three months. During this season the green feed of this range puts rapid gains and finish on the lambs both through the milk of the ewes and the grazing of the lambs. Especially is this true for early lambs with some mutton breeding.

Let's look again at this picture of the grazing use of the mountain lands of this region. Over the established migration routes, the herders trail the ewe and lamb bands from foothill spring range to the uplands, the spruce-fir and alpine zones, where the range forage is ready around July 1. The return migration, during September, precedes the first snows of the uplands; trucks loaded with the earliest and fattest of the lambs ease down the mountain roads for some weeks before.

Contrasted with this migration of sheep from the foothills to the high country, cattle usually work upward from the hills in a "drift" movement to the pine-bunchgrass zone, the lower mountain country rather than the high uplands. This natural drift of the cattle upward through the summer and downward in the fall — managed in some degree by fences, riding, and the location of salt — covers a period of four to six months.

More often than not the mountain rangelands lack sufficient

Mountain meadows, an important range type in both the ponderosa pine and the spruce-fir zones, afford good grazing. (U.S. Forest Service Photo)

Range feed in the spruce-fir zone of the Rocky Mountains is found in the "parks" and through the timber. (U.S. Forest Service Photo)

14

summer grazing capacity to balance the spring and fall ranges of the foothills and the winter feeds of the valleys. The result is a considerable overuse of the mountain rangelands. Nature soon raises warning signals when grazing of the mountain rangelands is heavy for any prolonged period. Bunchgrasses diminish and weeds increase, browse plants carry more of the grazing load, less accessible sites are used, easily accessible mountain meadows and parks take on a "peeled" appearance before the end of summer. If these warnings are not heeded, topsoils start moving on the slopes. When nature's bond of vegetation and mountain topsoil is broken, it may be too costly or too difficult to arrest the soil movement. Strangely, though, sleek cattle and fat lambs may come off these lands for a considerable period of years after nature's warning signals appear. On the mountain lands, somewhat in contrast with the plains grasslands, livestock production economics can lag behind a resource downtrend, but finally nature closes this gap with a vengeance!

Now, let's look at the ranches of this region. We see most of the cattle ranches located along the smaller valleys in the foothills and lower mountain country. Naturally productive ranches, these, with green range forage through the summer that puts bloom on

Sheep grazing in the "Land of the Short Summer," the alpine zone of the Rocky Mountains. (U.S. Forest Service Photo)

the calves and makes well-fleshed steers. Irrigated and subirrigated native hay meadows furnish added green forage in the fall after hay harvest for putting autumn gains on the animals to be marketed. We also see some cattle ranches located in the high mountain valleys. These ranches make beef the hard way, growing a great quantity of hay and feeding it to the cattle continuously during the five months or more of snow cover. Unique in many ways, these high mountain valley ranches can be very productive and profitable. Good illustrations of the high mountain valley cattle ranches are those of the Big Hole basin in southwestern Montana and of the North Park country of north central Colorado. In these high mountain valleys, more than 6000 feet in elevation, an ample harvest of good grass and sedge hay from flood-irrigated native hay meadows compensates for the short range season. This hay does a remarkable job of fattening older animals when fed in quantity through the winter. Often these ranches do such fattening, using only the hay, and market fat cows and steers in May.

We see most of the sheep ranches of this region in the larger and lower valleys where growing seasons are long enough for alfalfa hay production. Alfalfa, generally rated the best sheep hay, furnishes the protein needed for growing the wool. Two sheep ranch operating systems prevail, based on early and late lambing. Ranches with ample winter feed and a good upland summer range breed for March lambing, using sheds. These ranches use mutton-breed sires with fine-wool or crossbred ewes and market a large percentage of fat lambs from the summer range. Those ranches that do not have the natural advantages of good upland ranges and ample alfalfa production breed for range lambing, usually in May, and produce feeder lambs.

Intermountain Region. Let's turn our eyes now to the lands which lie between the Rocky Mountains and the Cascade and Sierra Mountain chains far to the west. This, the intermountain region, encompasses the great interior basin where no waters run to the sea, the plateau lands of the Colorado River drainage, the Snake River plains, and other parts of the Columbia River basin. (Refer again to Figure 1.)

"Arid climate" is written plainly over most of this region. Much of it is semidesert, with an annual precipitation of less than eight inches. Only the high mountains are subhumid.

Again we see the distinct native plant zones, considerably different from those of the Rocky Mountain region, but even more significant to ranch operations. In some places one may see all the natural zones of this region, from desert shrub to alpine, during a short drive up to and across one of the humid islands of mountain land. (A drive across the Grand Mesa from Grand Junction to Cedaredge, Colorado, affords this fascinating experience.) More often, though, we see the immense zones of the arid lands billowing to the far horizon. These, the large range types of the desert and semidesert country, require the long migrations which the sheepmen make in the use of seasonal ranges. Many range sheepmen make an annual trek of several hundred miles from the low desert winter range across intermediate spring-fall range to upland summer range and back to winter range!

For our view of rangeland natural zones in this region we'll start with the desert country. In this, the desert-shrub zone, we see the gray cover of the shadscale, saltbush, winterfat, and other plants of the desert-shrub association. This range type covers much of western Utah, central and eastern Nevada, and southeastern Oregon. Distinctly sheep winter ranges, these lands are used by migration from mountain summer rangelands such as the Wasatch highlands of Utah and the Ruby Mountains of northeastern Nevada.

Between the desert-shrub lands and the far-distant uplands we see a gradual transition to the sagebrush zone. This natural zone, semiarid rather than arid, occupies large areas of southern Idaho, northern Utah, northern Nevada, and eastern Oregon. Range technicians call this the sagebrush-grass range type because of the natural association of the big-sagebrush and the bunchgrasses. Sheepmen use this land as the spring and fall range in the annual movement between desert winter range and mountain summer range. These sagebrush lands also provide spring and early summer cattle range if they lie near the cattle ranches. Before the depletion of the bunchgrasses and the invasion of an annual bromegrass (cheatgrass) throughout most of this range type, these lands afforded much more spring and summer cattle grazing.

Western desert shrub, the main range type of the intermountain arid lands. (U.S. Forest Service Photo)

The intermountain sagebrush lands lie between the desert country and the humid islands of mountain lands. (U.S. Forest Service Photo)

Where the sagebrush lands merge with foothills and mountain slopes we see a zone of open woodland. In this, the piñon-juniper zone, the native plant association consists of piñon pine and juniper trees with an undergrowth of bunchgrass. This type occupies most of the foothills and lower mountain slopes in the region, also the lower of the Colorado plateaus. This land is cattle range, summer or spring and fall. Above the piñon-juniper lands there is usually a zone of oakbrush on the mountain lands of the interior basin and the Colorado plateaus. We see this zone most distinctly on the intermediate elevations of the Wasatch in Utah and on the plateaus of western Colorado and northern Arizona. This land is cattle summer range, with oak browse and bunchgrass providing most of the forage. (The use of oak browse is due mainly to the widespread depletion of the bunchgrasses in this range type.)

The ponderosa pine or pine-bunchgrass zone usually occurs between the 7000- and 9000-foot levels of the higher mountain lands in this region. Since the coarser of the bunchgrasses provide most of the range forage in this zone, cattle ranches use it as summer range. We see the spruce-fir and alpine zones on the uplands of the higher mountains; the natural conditions and grazing use of these zones in this region are very similar to those of the Rocky Mountains.

With this picture before us of the natural zones of these distinctly seasonal rangelands we take a brief look at the ranch operating systems of the intermountain region.

The desert sheep ranches usually have for their deeded land base some of the sagebrush land, the spring and fall range which lies between the desert country and the mountain land summer range. Much of the sagebrush lands and most of the desert-shrub lands are public lands, managed in federal grazing districts. Desert winter ranges provide the forage during the period from November through March, upland ranges from late June to mid-September, and sagebrush lands during the intervening spring and fall periods. These ranches use fine-wool sheep and market feeder lambs.

In contrast to the desert sheep operations, many of the Columbia basin sheep ranches feed hay through the winter, buying alfalfa hay from farms in the irrigated districts. Good lamb producers, these ranches use the mutton breeds and market early lambs.

A piñon-juniper zone occurs on much of the lower mountain country of the intermountain region. Seldom do we see this type of range in the good condition shown here. (U.S. Forest Service Photo)

Above the piñon-juniper lands in Utah and western Colorado we usually see a zone of oakbrush. The illustration shows this range type in good condition; overgrazing develops a dense stand of the oakbrush. (U.S. Forest Service Photo)

*Sometimes we see a definite aspen zone on the mountain lands of the inter-mountain region below the spruce-fir zone and merging with that zone.
(U.S. Forest Service Photo)*

Cattle ranches of the intermountain region do not make the extensive migrations that are typical of year-round sheep grazing. Usually we see the cattle ranches adjacent to or near the uplands, located along a stream bottom where hay meadows can be developed. Sagebrush lands provide the spring and fall grazing. Lack of upland summer grazing sometimes necessitates irrigated meadow grazing or summer use of the sagebrush lands. In some parts of this region desert or semidesert cattle ranches range their cattle all year long on desert-shrub lands that have some growth of bunchgrass. Operating at a low cost; with a correspondingly low production, these ranches are likely to be marginal.

Southwestern Region. Look southward now from the Tonto rim, at the country which lies between the Colorado River and the Rio Grande. Here we see that the rangelands and the ranches again differ from anything we have yet observed.

In this southwestern region of sunshine and *poco llueve* we see some startling contrasts between the creosote-bush deserts and the productive grama grasslands. Rainfall has two distinct seasonal "humps" in this country. Come the winter rains — *when* they

California annual grass type. Here we see this range type with a scattered growth of the California oak. At a lower elevation these grasses grow in a pure grassland type. At their upper elevation limits these annual grasses grow in the manzanita and ceanothus brush zone. (U.S. Forest Service Photo)

Southwestern desert shrub. Range forage consists mainly of winter-growing annuals. (U.S. Forest Service Photo)

come in volume — and the desert lands bloom in profusion with the winter annuals such as alfilaria and woolly Indian wheat. During the period from February through May these lands *sometimes* have a high grazing capacity. But the midsummer and late summer rains give the grama grasslands a much more dependable feed supply. These grasslands — most of them are in southeastern Arizona and southwestern New Mexico — lie between 4000 and 6000 feet. Another range type of some importance, the semidesert grass and brush type, lies intermediate between the low deserts and the grama grasslands of this region.

We see the best of the cattle ranches in the grama grasslands. Usually productive and stable, these ranches graze year-round with some winter supplement of protein concentrate. The cattle ranches of the semidesert lands, low in per-head cost and income, operate somewhat precariously. We do not see any ranches in the low desert lands. These lands are used through speculative inshipment of cattle when the winter and spring range season looks favorable.

Two kinds of range sheep operations prevail in the region. Early-lamb producers rent alfalfa pasturage in irrigated districts such as those of the Salt and Gila River valleys and market fat lambs in the spring. The dry ewe bands are then trailed to summer range in the national forests and Indian reservations of central Arizona, where they are grazed for a period of four to six months. Sheep operations that range year-round migrate between desert winter range and the high country of central and northern Arizona and New Mexico. These operations produce feeder lambs.

Pacific Region. We turn westward to the Sierras now for a look at California rangelands and ranching, remembering that we encompassed northeastern California, Modoc and Lassen counties, within the interior basin part of the intermountain region. For this region we shall confine our attention to California, since the Cascades and other mountains of western Oregon and Washington have mainly forest resources.

Far below the crest of the Sierras we see the valleys of the Sacramento and San Joaquin rivers merge to form the great central valley district of California. Far beyond the valley we see the dim outlines of the coastal mountains and hills (exceptional visibility, but it can be done!).

Southwestern semidesert grass and brush range type. These arid lands have probably been rendered more arid by overgrazing. (U.S. Forest Service Photo)

Southwestern mixed grama grassland, a productive range type of southeastern Arizona and southwestern New Mexico. (U.S. Forest Service Photo)

Around the central valley, on the tumbled lower foothills of the Sierras and the coastal hills and mountain slopes, the annual grasses of California spread their carpet. (Underline those words *annual grasses*; just about all the range grasses we've seen up to now have been perennials.) Lush and green during the season of winter rains, harsh and sear through the rainless summer, this range is unique in plant composition and season of growth. Annual bromegrasses and oatgrasses and annual weed forage such as the alfilaria grow in volume soon after the winter rains start during November. Shortly after the summer rainless season begins the annual grasses cure to a dry and rather poor feed. The alfilaria and other weed feeds vanish.

California cattle ranches in this zone of the annual grasses limit their year-round operation to the capacity of the range during the dry summer season. Then, to use the range capacity during the green feed season these ranches ship in large numbers of stocker cattle from Nevada, Arizona, and New Mexico for winter grazing and sale to packers or feeders in the spring.

Sheepmen that operate in the annual grass zone breed to lamb in autumn, selling the lambs in the spring when the range forage dries. They then move the ewe bands to high Sierra summer range or to irrigated land pasturage and crop aftermath in the valley. These sheepmen do considerable crossing of mutton breeds and fine-wool breeds, and they market a large percentage of fat lambs. Most of the lambs go to market in April and May.

Of some importance in California range livestock production, the rangelands and ranches of the upper foothills of the Sierras differ materially from those of the lower foothills. In the upper foothills the California annual grasses must compete with the plants of a brush zone. Here the manzanita and ceanothus often develop dense brush stands. These brush stands reduce the range forage growth, impede livestock and land management, and fray the good humor of the ranchman who must ride through them to work the cattle. Periodic burning of the brush stands increases the growth of the annual grasses for a time, but the brush regenerates quickly. Some controversial public questions of ranch management versus land conservation arise from this treatment of the rangelands.

Most of the ranches in this zone are small to medium cattle

ranches. In contrast with those of the lower foothills, cattle ranches in the upper foothill zone often move to the high Sierras for the limited amounts of cattle summer range available there.

California cattle ranches have a distinct price advantage because California beef consumption considerably exceeds production. This means large inshipments of cattle and processed beef. Beef prices of California central markets consequently approximate the prices at the inshipment markets — Ogden, Denver, and Phoenix — plus freight and handling costs from these points. This price advantage to California producers is probably permanent, since estimates indicate that in 1940, or before the wartime growth of California cities, beef consumption was about double the state's production.

We've now swept our view across the ranch country west of the hundredth to see the immense pattern of western lands and ranches. Next we are going to have a closer look at the ranches pictured briefly in this panorama, examining the main types of ranches in each of the natural regions. Always interesting, often fascinating, western ranches mirror the many contrasts of nature, reflect the many different ways the ranch country is "put together."

WESTERN RANCHES

Northern Plains Ranches. Between 1905 and 1920 a period of rapid homestead settlement brought most of the northern plains lands into private ownership, the Indian reservation lands excepted. This development modified ranch operations considerably, and one may see clearly the influence of this period of rapid homesteading on some of the types of ranches that now prevail. But in spite of this rapid homesteading of the northern plains, only a minor percentage of the total land area was farmed; probably less than 10 percent of the total area was ever plowed. Crop agriculture (mainly spring wheat) now occupies considerable areas of the rolling glacial plains north of the Missouri River, but it accounts for only a small part of the northern plains roughlands that lie between the Yellowstone and the North Platte rivers.

One type of northern plains ranch bearing a distinct imprint of the homestead settlement period is the small cattle ranch which has evolved from the original homestead farm. This type of cattle ranch, numerically the most prevalent in the northern plains, has

50 to 150 head of cattle, owns two to four sections of land, and leases an equal amount. This, the family-size ranch of the northern plains, is a fairly self-contained unit. Most of these ranches grow their own winter feed, and the operator and his family do most of their own work.

By other western ranch standards these ranches represent rather intensive range cattle production. The livestock investment consists mainly of the beef cow breeding herd; the feeder calves make up most of the marketing. For these ranches the trend — which is likely to continue — is to retain part of the calf crop for feeding on the ranch to make a slaughter animal at about nineteen months of age. There are several different variations in this feeding practice, depending on the feed resources of the ranch. For illustration, one ranch retains half or more of the calf crop, gives the calves a grain supplement through the winter, ranges them through the summer, and then gives them access to standing dryland corn during their second autumn. In a good crop season this ranch makes slaughter animals weighing about 1000 pounds at the end of the second autumn (for calves born in April). The operator of this ranch says that he markets "beevetes."

Another type of northern plains cattle ranch, *not* greatly changed by the homestead period, is that based upon the early ownership of a considerable body of land. These ranches have been built by the business acumen of the operator, the acquisition of lands from homestead owners (often the ranch cowboys "homesteaded," too!), and the ownership of key lands and water to control leases. Outside investment capital has been an important factor in the development of some of these larger ranches. Often this type of ranch is a corporate form of business organization, though the shares are usually held in the family or by a small number of people.

This type of ranch, the large cattle ranch (from 500 to several thousand head of cattle), accounts for a fairly important part of northern plains range cattle production. These ranches do not, as a rule, excel in their production per head or in any other unit measure of production efficiency. Their strength lies in having a business of sufficient size to afford a larger margin for meeting risk and for keeping abreast with technological progress.

Often this type of ranch combines a range cattle enterprise with some dry-land farming or with a range sheep enterprise. These two range livestock enterprises complement each other where both cattle range and sheep range (or dual-use range) are an important part of the rangelands of the ranch, or where some of the feed crops grown on the ranch have a special suitability for one or the other kind of livestock. However, range cattle and range sheep enterprises on the same ranch are not complementary to any extent in the use of labor time for the management of the livestock. As a consequence, each enterprise needs to be large enough to afford adequate employment for the specialized labor force. For these northern plains ranches this minimum size is around 150 head of cattle and 1200 ewes.

Cattle ranches of the Nebraska sandhills constitute a distinct type, even as the sandhills appear as a distinct area in the northern plains. These ranches, medium to large in size, operate primarily on a breeding herd basis. Except on the fringes of the sandhills these ranches were not much disturbed by any rapid homestead settlement. The reasons? The homestead settler shunned this land where plowed soils blew away and where the sand-point well often had to be driven a hundred feet for water. Then, the early cattlemen of the sandhills found the coarse grasses quite deficient in protein, for both winter range and native hay.

In comparatively recent years this has been considerably changed by the advent of high protein concentrates for range and hay supplements. Cottonseed and soybean meal in "cake" or other form has been a real boon to sandhills cattle production. Only twenty-five years ago the Nebraska sandhills were a land of low values, low production, low income per unit — an uninviting prospect for any considerable development of small ranches. Public expenditures and taxes for roads, schools, and other public services are low in this area, but this, though conducive to a low land cost for the large ranches, does not attract the development of small ranches.

Two main types of sheep ranches prevail in the northern plains. One is the family-size ranch operating with 1000 to 3000 ewes; the other is the corporation ranch with Indian reservation or other public land leases, operating with 5000 to 30,000 ewes. A few of these large ranches own considerable land.

Family-type sheep ranches own haylands and croplands, usually irrigated lands along the larger streams, and own and lease adjacent rangelands. Some of these ranches depend on dry-land crops, usually grain hay, for winter feed. The larger ranches with Indian land or other public land leases move the sheep considerable distances between their ranch properties and the rangeland leases or between summer range and winter range. Some move to winter feed in the irrigated farm districts where they buy hay and other crop feeds. As a rule, these ranches operate at a low cost, but they are hazardous in their limited provision for winter feeds.

Most range sheep operations in the northern plains have in the past used a type of the Rambouillet sheep breed tending to a fine-wool fleece. In recent years these ranches have shifted rather strongly to the Columbia breed in an attempt to improve their lamb production without appreciably sacrificing wool quality.

Ranches South from the Platte. Stock farms, one to four sections of land in size, characterize much of the livestock production in the central plains region. These stock farms grow dry-land forage crops (grain sorghum, corn, wheat hay), have some unplowed native pasture land, and derive most of their income from beef cattle. Many of these stock farms grow some cash grain.

In some parts of the central plains cash grain farming predominates almost to the exclusion of range and pasture livestock production. Crop agriculture occupies much more of the central plains than of the northern plains. Large areas of western Kansas, eastern Colorado, and the Texas Panhandle are now devoted to fall-sown wheat production, with minor livestock enterprises on the farms. From this cash-crop agriculture we see transitions in other localities to stock farms where the accent is on livestock and feed crops.

In some parts of the central plains soils and topography forbid crop agriculture. Several such areas — stock ranching areas — are of considerable importance; for example, the Loess Hills country of the Republican River drainage in southwestern Nebraska and the broken topography of the Purgatoire River drainage in southeastern Colorado. Then there are some areas of the central plains where unwise crop agriculture has left broken and sick lands, lands where some semblance of a range plant cover now struggles for a foothold.

Short grasses, principally buffalograss and the grama grasses, predominate in the native grassland pastures used by the central plains stock farms. These grasses by themselves do not afford a well-balanced and season-long pasturage — they are essentially summer grasses. As a consequence many of the central plains stock farms use these native pastures as a supplement to the crop feeds and pasturage of the cultivated lands. Often one sees these native pastures severely overgrazed by the end of summer. Fortunately, these short, sod-forming grasses are amazingly tough; they stay on the job as soil protectors even though overused until the volume of range forage production is severely impaired. But, eventually, these pastures break under such use. Frequently they are grazed at a rate of one to one and a half acres per animal month of use (cattle) where a rate of two to four acres per animal month would produce more total livestock gains and a better market grade for cattle.

Some complement of the midgrasses originally constituted an important part of the central plains range plant association. Grazing pressure diminishes the cool-climate midgrasses more quickly in the central plains than in the northern plains. Good range and pasture management, however, can retain and restore the midgrasses in many localities of the central plains. These grasses and many of the native weeds and shrubs frequently afford an important addition to the feed value of the native short grasses, and they also help to balance the summer and early fall grazing capacity of the short grasses with spring and late fall grazing. Absence of the midgrasses in association with the predominant short grasses of this region usually may be taken as an indication that a central plains range is used beyond the optimum rate.

To a considerable extent the stock farms of the central plains function as intermediaries between the western range feeder-livestock grower and the midwestern feedlot feeder in the production of finished beef cattle. Central plains stock farms buy feeder cattle from the southwestern and Rocky Mountain region ranches. These cattle are fed to a partial finish on the surplus crop feeds of the central plains stock farms and then shipped to the markets at Omaha, Kansas City, Chicago, St. Joseph, and Wichita. From these markets a considerable number of these cattle are purchased by the feedlot feeders of the Corn Belt.

We look now at the ranches of the southern plains. We saw few sheep ranches in the central plains region; we see many of them in the southern plains. Here sheep numbers continued to rise through the recent war period, although the national sheep population declined sharply. Texas alone had, in 1945, approximately one fourth of the total national population of stock sheep on farms and ranches. The ample local supply of Mexican labor and the lower operating costs of the southern plains sheep ranches evidently accounts for their ability to move counter to the national trend during the war period.

Three major areas, all of them geographically immense, make up the ranching parts of the southern plains region. Most important of these is the Edwards plateau, which occupies west central Texas east of the Pecos River and extends nearly as far east as Austin, Texas. Del Rio and San Angelo, Texas, mark the approximate southern and northern limits of the Edwards plateau. Next most important in southern plains ranching is the Rio Grande Plain district of Texas, which extends southeastward from the Edwards plateau to the gulf, between the Rio Grande and the Guadalupe River. Finally there is the Trans-Pecos district, which includes the part of Texas west of the Pecos River and the southeastern part of New Mexico.

In the Edwards plateau, the great sheep-ranching empire of west central Texas, we see the type of sheep ranch that characterizes much of the southern plains sheep ranching. These Edwards plateau sheep ranches, usually of medium size and with a considerable investment in rangelands and improvements, range the sheep in large, fenced paddocks. Typically these ranches have beef cattle and often mohair goats supplementing the range sheep enterprise. The type of range, which is a mixture of grass, weeds, and browse, favors this combination of range livestock enterprises. (However, such use of this range requires unusually skillful management to avoid shifts in the range type and consequent changes in the balance of the capacities for the different kinds of livestock. Such shifts have been induced on a considerable scale in the western and more arid parts of the Edwards plateau.) These sheep ranches use the Rambouillet breed, tending to a fine-wool type, and market feeder lambs. They also sell some yearling wether sheep to the limited slaughter markets for this class of animal.

Cattle ranches, some of them very large, predominate in the Rio

Grande Plain district of Texas. We see but few sheep in these warm and subhumid lowlands. Here we find the greatest use of the type of cattle resulting from a cross of the Brahman or Zebu with one of the American beef breeds, usually with the Shorthorn. These crossbred animals, more resistant to the parasites and better adapted to the heat than the Hereford, have come into general use here during the past two decades.

In the Rio Grande Plain we see in its extreme form the invasion of rangelands by the mesquite tree. Range forage growth is often severely reduced as a result of such an invasion, but the removal of the mesquite growth is very costly. Mechanical removal of the mesquite and reseeding of the range have been undertaken on a considerable scale by some ranches, especially on rangelands that were of high grazing capacity before the mesquite invasion. Some of these lands are capable of carrying one head of cattle through the year for each six to twelve acres of land. Some of the exotic grasses that have been used in the reseeding attempts show a favorable response to the soils and the climate of this district.

In the arid Trans-Pecos district of the southern plains we see principally cattle ranches. These semidesert and desert grasslands have been greatly changed by heavy grazing use, and the economy of the ranches has been consequently affected. Rangeland grazing capacities have declined sharply with the diminution of the perennial grasses, the increase of the annuals, cacti, and juniper, and the loss of range soils. As a consequence we see a trend on the cattle ranches to change from a breeding herd and the marketing of feeder calves to the operation of a mixed herd and the sale of thin stocker cattle. This is, of course, a decrease in productivity and a lowering of the economy of the ranches. There is also some shift to sheep ranching, in order to use the weeds and the browse of the changed range type.

Valley and Foothill Ranches. We see three main types of cattle ranches in the Rocky Mountain region — foothill ranches, lower mountain valley ranches, and high mountain valley ranches.

Foothill ranches, the most numerous of the three types, operate on foothill grasslands. As a rule, this rangeland is all or nearly all owned by the ranch. These ranches use ten to twenty acres of rangeland per head for the range season of about eight months, grow irrigated native hay or alfalfa for the winter feed, and often produce some small grain for supplemental feed. These ranches have in the past marketed principally yearling and two-year-old steers, though recent high prices for feeder calves have induced the sale of more calves and an increase in breeding herds. Many of these ranches are able to grow long two-year-old steers to a weight of 1000 pounds at thirty months of age — well-fleshed animals that require little feedlot finishing. Though the Rocky Mountain foothill cattle ranches vary greatly in size, the mode is fairly large, probably about 400 or 500 head of cattle. These ranches, productive and stable, achieve good production with moderate operating costs.

Cattle ranches in the lower valleys of the Rocky Mountain region depend to a great extent on the use of irrigated lands. There appears to be a trend for many of these lower valley ranches to operate the year round on their irrigated crop and pasture lands, fitting in with the development of modern types of beef animals better suited to pasture than to upland ranges. Feeder calves and yearlings constitute the principal marketing of this type of ranch.

These ranches of the lower valleys often own some adjacent foothill rangeland for spring and fall range and have a national forest summer range permit. Like the foothill ranches, they have a high land investment, since cropland feeds provide five months or more of the year's feed. Of the several types of western cattle ranches, these ranches are the highest cost operations and have the highest unit production. For this type of cattle ranch, good management is a necessity and poor management is quickly and severely penalized.

Cattle ranches of the higher valleys (above 6000 feet) of the Rocky Mountain region must, as a rule, depend on large acreages of irrigated native meadows to produce the five or more months of winter feed supply. These high valleys lie above the climatic limit of crop agriculture, except in the southern Rocky Mountains, and the growing season is usually too short for alfalfa hay. Native hay meadows are flood-irrigated during the spring, and the one cutting of hay is harvested in August.

These ranches typically have a natural overbalance of hay-producing land in relation to the limited summer range for cattle above the pine-bunchgrass zone of the mountains. Often the ranches compensate for this unbalance by using some of the meadowland for pasturage and by purchasing in the fall cattle from outside the valley for winter feeding on hay. Usually the native hay of the high valley has a good fattening quality, because of the high proportion of good sedges. An ample feed of this hay through the winter will produce a fair finish on cows and steers. Feeder-calf sale should be, and usually is, minor in the marketings of these ranches.

We seldom see sheep ranches located in the high mountain valleys of the Rocky Mountain region. They are mostly in the lower valleys where crop feeds and tame hay can be grown. The sheep ranches that breed for early lambing and range-fat lamb production usually have a considerable investment in lambing sheds and feed ample amounts of hay and grain before and during lambing. These ranches generally breed for March lambing. Mutton-breed sires, Suffolk, Hampshire, or Shropshire, are used with Rambouillet or Rambouillet and mutton-breed crossbred ewes. These ranches must have a good upland summer range to make range-fat lambs weighing 85 to 95 pounds by late summer. These are the high-cost sheep ranches of the West; they must have good lamb production to succeed. In size these ranches vary from one to five or more summer bands of 1000 ewes each. The modal size for this type of sheep ranch appears to be about three bands totaling 3000 to 3600 sheep. Typically, these ranches own fifteen to twenty acres of irrigated hayland per 100 ewes, own and lease three to five acres of foothill spring-fall range per ewe, and have a national forest summer grazing permit for the ewe and lamb bands.

Compared with the early-lambing sheep ranches of the Rocky Mountain region the range-lambing sheep ranches — the other main type — have a considerably lower investment in land and improvements. They also have a much lower production cost, a lower production, and lower incomes. They use the Rambouillet breed of sheep, and wool sales contribute a much higher percentage of the gross revenue than for the early-lambing ranches. An important exception to this observation should be noted, however, in some of the Colorado sheep ranches on the west slope of the Rocky Mountains. These ranches have unusually favorable summer ranges and produce range-fat lambs with May lambing. These operations make considerable use of the mutton breeds. However, most of the Rocky Mountain region range-lambing ranches pro-

duce a feeder lamb weighing sixty to seventy pounds when marketed in October. Many of these ranches do not have high-country
summer range — a good reason for not undertaking range-fat lamb
production.

Ranches of the Arid Lands. Climate in the intermountain region varies from semiarid to extremely arid. In some localities the
capacities of the seasonal ranges were not naturally well balanced.
Again some of the naturally well-balanced seasonal range capacity
relationships have been unbalanced by overgrazing and range depletion. To a large degree the method of analyzing ranch management in this region must be in terms of the resource problems
of the ranches. Arid lands have been made more arid. Sweeping
changes wrought upon the resources have resulted in sweeping
changes in the ranch operations. For illustration, the sagebrush
lands once afforded much season-long range for cattle. The sagebrush stands were open, the bunchgrasses were an important part
of the ground cover. Now, in contrast, the native grasses are scant,
the sagebrush heavy, the cheatgrass usually produces green feed
in the spring but is harsh and beardy by midsummer. This range
now affords good spring and fall sheep grazing and is so used when
located between winter rangelands or winter feed-crop areas and
mountain summer ranges.

This change in the sagebrush lands has profoundly modified
many of the cattle ranches that were located in this range type.
The number of cattle that can be maintained by these ranches is
sharply down; the irrigated hay meadows that these ranches have
developed along the streams now provide much of the cattle grazing past midsummer. Hay meadow aftermath and regrowth now
substitute to a large extent for summer range, and "The cows can
hear the rattle of the hay mower calling them in off the range miles
away." Typical illustrations of the changes here described are the
Harney basin of eastern Oregon and the Ruby valley of northeastern Nevada.

For a further and closer view of the ranches of this region,
let's again look first at the desert sheep ranches. These operations
use the desert areas such as the Red desert of southwestern Wyoming and the Utah "west desert" (the arid lands west of the Wasatch highlands) as their winter range from early November to
early April. They then migrate to sagebrush range for spring graz-

ing and for lambing, en route to the summer ranges. Some range through the summer season in the higher elevations of the sagebrush lands, but with adverse effects upon lamb production. These desert operations use the Rambouillet, usually a small, hardy, and fine-wool type of this breed. The average weight of the feeder lambs falls usually between 55 and 60 pounds. These are low-cost sheep ranch operations with a correspondingly low production and low income. Their margin of net income can be satisfactory with good management. The winter ranges used by these ranches were, before the enactment of the Taylor Act in 1934, the largely open and uncontrolled public domain lands. These winter ranges have deteriorated markedly. A winter season's supplemental feeding of 20 to 30 pounds of protein concentrate per ewe is not unusual.

Another important type of sheep ranch of the intermountain region is exemplified by the sheep ranches of the Snake River valley in southern Idaho and southeastern Oregon. These operations are unusual in that they own irrigated croplands in irrigated farming districts. They usually buy some of their crop aftermath fall pasturage and winter feeds from the nearby farms. As a rule these ranch operations lamb early, using sheds, move to the adjacent sagebrush lands of the Snake River plains for spring range as soon as the cheatgrass affords green feed (late March), and then migrate to summer ranges in central Idaho.

These ranches use large amounts of hay and grain through the winter and during lambing. They also buy crop aftermath pasturage for the ewes in the autumn, especially when the fall range regrowth of cheatgrass and weeds is poor. Records indicate that this type of sheep ranch excels all others of the West in lamb production. It also has the highest operating costs. Range-fat lambs are marketed, mainly in August. The lambs are three or more months old when they go onto high summer ranges, and the greatest range forage value of these high summer ranges comes from their use as finishing ranges for the lambs. Unfortunately, the ewes are held on these ranges too late in the fall, and the consequent overuse of the high country interferes with the use of the summer range as a finishing range for fat-lamb production.

Cattle ranches of the intermountain region fall geographically into two main types. Those of northern Utah, southern Idaho, eastern Oregon, and northeastern Nevada have irrigated feed-crop lands as the base for the use of nearby seasonal rangelands. These ranches own considerable rangeland, especially spring-fall and summer range. Irrigated meadow and pasture grazing often constitutes an important part of the ranch capacity. The cattle ranches of the Bear valley of southeastern Idaho and northern Utah are a good illustration of this. Because of the influence of the West Coast cattle markets, these ranches aim to produce "two-way" beef cattle for these markets from their ranges and pastures. These are mature animals, as well fleshed as possible from grass. These cattle go to the California markets for slaughter or for a short feedlot finish. The recent trend for these ranches has probably been to sell more yearlings to intermountain and West Coast feeders.

For southern Utah, northern Arizona, and central and southern Nevada the cattle ranches may be described as semidesert, year-long range operations. Their main land holding is the rangeland with the water which has been or can be developed to provide the water service for the adjacent public domain range. These ranches have, as a rule, a low unit production, low income, and low operating costs. These operations are likely to be hazardous for the family ranch, and they tend to be fairly large and somewhat speculative. In most seasons their marketings are thin steers and cows. These go to feeders for a short feed, and they are then marketed as "medium" to "common" beef animals.

We made the observation that many of the arid lands of the intermountain region have been made more arid because range plant and soil depletion has decreased the effectiveness of the moisture. This same observation applies to much of the southwestern region, excepting the grama grasslands of southeastern Arizona and southwestern New Mexico. The creosote-brush desert lands, now worthless except during the seasons when the winter annuals grow in volume, once supported a fair amount of perennial range vegetation. Many of the semidesert lands that once bore the aspect of grasslands have become a thorny tangle of brush. One must interpret many of the present ranch operations in the light of these changes in the resource.

Most prevalent as a ranch type in the southwestern region is the semidesert cattle ranch. These ranches can be fairly productive. Many of them succeed rather well as a family-size operation;

others were made precarious when the grasses were diminished
and the mesquite brush and the catclaw closed in. These ranch
operations range all year long; the better ones produce "good"
grade feeder calves at a low cost. Where brushy ranges impede
management, the marketings are likely to be mature animals, low
in quality and poor in flesh. This type, the southwestern semidesert cattle ranch, usually has about as much investment in the
water development (wells and windmills or power) as in the
rangelands. Leased state lands are an important feature of the
land setup of these ranches, since Arizona and New Mexico received a federal grant of four sections rather than two sections per
township.

Cattle ranches in the grama grasslands of southeastern Arizona
and southwestern New Mexico (these lands are a part of the
"Mexican highlands" physiographic province) are productive and
moderate-cost operations. Their range growth volume fluctuates
more from one season to the next than would a range with a more
variegated plant association. There is the probability that this situation is due partly to the diminution, under grazing pressure, of
some palatable weeds and shrubs that formed a natural complement of the grama grasses. Since the principal growth period of
the grama grasses of this range is August and September, the
ranch operator knows fairly well at the time of marketing, in October, the amount of the range forage available for the year ahead.
He can adjust his marketing and the size of his herd accordingly.

The principal sheep ranch type of this region is migratory, usually having but little in the way of a ranch property base. Desert
and semidesert public domain lands of southern Arizona and western New Mexico constitute the fall, winter, and spring ranges, the
national forest lands above the Mogollon rim provide the summer
range. These ranches use the fine-wool Rambouillet sheep that
most of the western desert range sheep operations use. The lambs
are feeders.

Alfalfa pasturage leased from the farms of the irrigated districts
of the Salt and Gila River valleys furnishes an important part of
the feed for the other sheep ranch operating system of this region.
The ranches using this system lamb on the public domain lands in
autumn, move to the irrigated pasturage in November or December, and market spring lambs in April. These operations are cost-

ly. Before the recent period of high wartime prices these operators paid thirty to forty cents a month per ewe and lamb for the irrigated alfalfa pasturage. This compares with the rangeland rental price, usual at that time, of six to eight cents per ewe month. High lamb production and the favorable seasonal price for spring lambs compensate for the high cost. After the sale of the lambs the dry ewe bands are trailed to some of the higher elevations of the public domain lands or to national forest ranges for the period from May to October.

Unusual Ranches, Strange Lands. California cattle ranches in the open grasslands of the annual grass type have a strange unbalance. They have ample green range forage from November through May. But from June through October the cattle lose weight on the dry range unless they are given supplemental feed. In increasing numbers the cattle ranches in the California annual grass range type now supplement their summer grazing with cottonseed cake or other high protein feeds. Probably before the time the waxing herds of the California dons reached their zenith on these lands, several species of perennial grasses gave this range a much more balanced capacity through the year, as indicated by some relic stands of the original range type, but the truth regarding this is now pretty much an academic problem. Anything that may in the future be done to put usable summer feed on these ranges will no doubt be through artificial means, not through restoration of the original type.

Many of the cattle ranches in the lower Sierra foothills buy stocker cattle in the autumn for winter pasturage and spring marketing. These are grazed through the winter green feed period as an addition to the regular herd maintained through the year. Some ranches are able to increase the size of their regular herd by moving part of the herd to leased irrigated pasturage in the central valley for summer grazing. This is an alternative to the purchase of stocker cattle in the fall for winter grazing and spring marketing. These lower foothill cattle ranches seldom move to the Sierra uplands for summer range. The distance is considerable, and the U.S. Forest Service policy has favored the smaller cattle ranches of the upper foothills in granting permits for the limited amount of upland cattle range in the national forests of the Sierras.

California range sheep operations also contrast rather startling-ly with most of the other western sheep ranches. To a surprising degree they are based upon the use, through purchase and lease, of the immense reservoir of crop feeds and aftermath and irrigated pasturage of the central valley. These ranches are usually located in the lower foothills, and the annual grassland range affords the green winter feed for growing the lambs. Husbandry practices are aimed at making a range-fat lamb by April or May. These ranches have done considerable crossbreeding of fine-wool and mutton-breed ewes in the attempt to increase lamb production without changing entirely to a mutton-breed ewe. Often this crossbreeding has not been too well planned for the production of uniform wool and a good quality fleece, with a resulting loss in wool income.

California ranch properties and rangelands carry some very high capital values, second only to those of Texas. California cattle ranches before the war often had an investment in land and ranch improvements of $100 to $120 per head of cattle. This was caused in part by the favorable market price situation which these ranches had. Since California has for some time been a net importer of lamb, as well as beef, this price situation also favors the sheep ranch incomes and capital values. However, many of the higher investments in ranch property *per unit of capacity* are found on the poorer ranches, where the productivity and capacity of the ranch has been misjudged. Our observant California ranch operator commented in 1940, before the recent price distortion of ranch property, that "The *minimum* price for rangeland seems to be five dollars an acre in California."

RANCH PROBLEMS

We have seen the natural-region pattern of resources and ranches in the West. We have observed the great variations in resources and the many consequent contrasts in the way ranches operate. We now look at some of the ranch management problems directly related to the land resources of the ranches.

Balancing the Seasonal Capacities. In their adaptation to natural factors ranches seek a well-balanced seasonal relationship of range forage and feed-crop capacities. This, a "balanced ranch unit," is *the basic* management requirement of all western stock ranches. Throughout the West since about 1920 stock ranches have increasingly used hay and other feed crops, pasturage, and protein concentrate range supplements to attain a balance in their capacities through the seasons. By 1920 most ranches had developed the necessary hay and feed-crop production for their winter feed season. Much, though not all, of the increased production and use of crop feeds, pasturage, and range supplements by ranches since 1920 has resulted from some unfavorable trend in their rangeland capacities.

We've already had some glimpses of this ranch problem. Now let's observe more closely, for some of the main types of ranches of each region, the natural balance in seasonal capacities, the unbalances, and the attempts of the ranches to offset their unbalances. We also want to see the influence of this upon the economy of the ranches.

Northern plains ranches attain a natural balance through the range season for their type of range by having a good complement of the midgrasses and the short grasses. Short grasses provide the late summer and early autumn range requirements, midgrasses the spring and early summer and the late fall grazing. Western wheatgrass, the most important of the northern plains midgrasses, is naturally deficient on some ranches because they lack a sufficient number of the favored moisture sites (swales, north slopes, drainageways) where this grass grows best. These ranches can usually offset this deficiency with crested wheatgrass seeding, which succeeds on drier sites than the native western wheatgrass and gives equally good results (sometimes better) in producing spring and fall grazing and range hay. Usually the northern plains ranches can produce ample winter feed to balance the rangeland capacity. These feeds — alfalfa produced with spring runoff irrigation, small-grain hay and straw, sweet clover, native hay — are needed in some years in considerable volume, in other years hardly at all.

Of all regions the ranches of the northern plains probably have the least problem of unbalance caused by changes wrought upon the rangelands through overuse. Rather their problem appears to be extreme climatic fluctuation. The great drouth seared the northern plains during the thirties; the years from 1939 through 1948 were phenomenally good. The drouth, so long and so intense that it appears climatically episodic, could not be coped with by the best management. However, the ranches that were overusing their ranges before the drouth have witnessed a relatively slower recovery of the midgrasses than of the short grasses. Those ranches do not yet have sufficient spring and fall range forage to balance the summer growth of the short grasses.

Our observations regarding the complementary relationships of midgrasses and short grasses in affording a seasonal capacity balance for northern plains rangelands also apply to much of the central plains. The economy of the central plains stock farms is such, however, as to lessen the importance of midgrasses in the use of the native grassland pastures by the stock farms. Their main interest is to get the summer grazing furnished by the buffalograss and the grama grasses. But this is not true for the ranches, since they need season-long grazing; and many of the central plains ranches have seen their range slip from a season-long range to a short-grass summer range because overuse took out the midgrasses. These ranches are then handicapped in operating on a breeding herd basis. Some of them have of necessity changed from a breeding herd operation to the buying of yearlings in the spring for summer pasturage and resale in the fall. A good many of these lands would recover their natural growth of midgrasses and useful shrubs and weeds if given a few years of nonuse or light use. Ranches using these lands would then again be able to function as year-round operations. Crested wheatgrass can in some localities of the central plains be used to supplement the depleted midgrasses.

There have been two significant indications of changes in southern plains range resources. One of these is the large and continuous rise from 1920 to 1940 in the sheep population of Texas and southeastern New Mexico. The other is the fact that by 1940 Texas was a net importer of cottonseed meal and products, even with the large production of these in Texas. Bearing out these indications, there has been an observed shift of some grassland ranges to an increasing amount of brush, weeds, and cacti. This change in some of the grasslands, apparently induced by prolonged overuse by cattle, has been accompanied by an increased use of protein supplements in the winter season (often an economic practice with moderate grazing) and by the addition of sheep and goat enterprises to use the increased brush and weed feed. Then, in the extreme cases, as the brush closed in on formerly productive grasslands, the brush and mesquite and juniper trees have been grubbed, cut with power saws, or otherwise removed at considerable expense preparatory to range reseeding. When their rangelands decline to this condition, ranches cannot compete with well-managed range and pasture livestock production.

We have observed that, in contrast to the ranches of the plains regions, the ranches of the Rocky Mountain region must fit some distinctly different seasonal ranges into their land setup. These ranches usually have a spring-fall range, a summer range, and croplands which provide winter feed and some fall pasturage. Here the resource management problem is to attain and maintain a reasonable capacity balance for these three kinds of land. A deterioration in any one of these seasonal lands unbalances the ranch.

Although an expansion of the feed-crop production to balance the capacity of the spring-fall and summer ranges was once a real need for many of the ranches of the Rocky Mountain region, such production is now usually adequate. As a general rule, ranches of this region are now more likely to have an overbalance than a deficiency of winter feed capacity. This has been caused by the continuous development of haylands by many ranches throughout the region, the increased use by ranches of concentrate feeds (corn, cottonseed "cake," commercially prepared "pellets" for range feeding) and a slow though sometimes considerable decline in the capacity of the rangelands.

Stock ranches of the Rocky Mountain region generally maintain the livestock through all or part of the winter months on hay and other crop feeds. There are some localities where four or five months of continuous winter feeding are a necessity; in others only intermittent winter feeding is necessary, and the hay is a supplement to the winter grazing. The capacity of the ranch for winter feed production should match the capacity of the grazing lands for grazing livestock through the average range season, or exceed it somewhat, to provide for some accumulation of winter feed reserves.

Hay is a high-cost winter feed, three to four times that of grass where there is some margin of choice between winter grazing and hay feeding and where the livestock production results from wintering on grass and on hay are somewhere near comparable. There are many ranches and some localities where the period of winter maintenance on hay and other crop feeds has been extended, especially into the spring season, to compensate for a decline in the capacity of the ranges, although better development and management of the grazing lands would generally be a more effective and cheaper means of attaining balance.

Like the ranches of the Rocky Mountain region, the ranches of the intermountain region seek to attain a balanced operation by fitting distinctly seasonal ranges into the ranch unit. The change that has taken place in the sagebrush lands, the main spring and fall range type of these ranches, causes their greatest problem in attaining a balanced operation. This change, one of the most extensive and complete that we see in a large range type, has, during the past fifty years, had a far-reaching effect upon the operations of the ranches.

With the depletion of the bluebunch wheatgrass, slender wheatgrass, and Sandberg bluegrass of these sagebrush lands, the sagebrush thickened to a close stand, and the cheatgrass replaced the native bunchgrasses. Eventually this range type no longer afforded season-long cattle grazing, the former rather general use of these lands. Those cattle ranches that could do so reorganized their operations more and more around their native meadows, but this meant fewer cattle and a higher per-head operating cost. Other ranches with small meadowland resources could not make such a change and they became bankrupt because their resource was bankrupt. In some such instances the real reason for bankruptcy was not suspected by the operator of the ranch. Overcapitalization, credit stringency, drouth, were sometimes thought to be the real reasons for ranch bankruptcy, when they actually were the secondary and superficial reasons.

The effects of this change in the sagebrush lands upon the ranches may be read further in the statistics of the cattle population of some parts of Utah, Nevada, and Oregon. In these instances the county statistics of the beef cattle population, where not obscured by developments in farming communities, show a peak between 1910 and 1920 followed by a drastic decline from which there has been no appreciable recovery. The sagebrush range type of the intermountain region, immense and potentially productive, offers an opportunity and a challenge to management.

This change in the sagebrush lands also unbalanced the sheep ranches, but they were able to reorganize their operations more successfully. There are sheep ranches in the intermountain region that now maintain the ewe bands on hay through a considerable part of the winter, although the ranches formerly did very little winter feeding. They used the sagebrush lands through the fall, winter, and spring months and then migrated to mountain summer range. As the cheatgrass replaced the bunchgrasses on the sagebrush lands these lands afforded more sheep spring range, less winter range. During the time this change occurred, mainly between 1890 and 1910, irrigation farming was expanding and affording more fall and winter feeds. As a consequence the sheep ranches were able to shift from range lambing to early lambing, from winter range to crop feeds, and the cheatgrass provided (usually) more and earlier spring forage than did the native bunchgrasses. Many of the sheep ranches now operate at a higher economy than formerly.

We turn now to a view of the problem of capacity balance for ranches of the southwestern region. The cattle ranches in the grama grasslands, the "southwestern mixed-grama" type, graze throughout the year on these same rangelands. Their problem of seasonal balance in range capacities is, consequently, similar to that of the plains ranches. The range needs such management as will give the best attainable growth relationship among the range plants providing the forage through the different seasons. Since the sheep ranches of this region use seasonal rangelands and migrate considerably, their problem of attaining a well-balanced capacity is comparable to that of the intermountain sheep ranches. Any considerable change in the forage capacity of any one of the seasonal lands used by a range sheep operation unbalances the year-round capacity of the operation. It seems probable that many of the southwestern range sheep operations were none too well balanced in their early period, and that the changes which have occurred in the desert and semidesert lands have accentuated this problem.

Resource problems of the California central valley foothill ranches would be much simplified if their rangelands had summer grazing capacity to match that of the other seasons. Perhaps eventually grasses can be introduced on these lands to increase their summer grazing capacity. If this is to be done, a much better comprehension of the ecology of the resource will be needed. An ecologic study of this main California range type may give an understanding of the changes that have occurred in it and may point the way for introducing into this range type some perennial range forage plants that will endure grazing and a hot, arid summer.

Where the natural balance of range plant life is undergoing adverse change, as is the case with many ranches and rangelands, the application of the science of ecology and the work of the practicing ecologist is needed now while there is yet time to avoid the thing that has happened to what is now the California annual grassland type. An ecologic analysis of a range plant association can give answers on what is happening to it and why, and it can make practical recommendations on changes in seasons of use, kind and numbers of livestock, reseeding, use of feed supplements, and the like. The use of this service by the ranchman, who often in the past has not comprehended the changes occurring in a range, can avoid a bankruptcy of the resource.

Adapting Ranches to Resources. By now the ranches of the West have generally fitted their type of operation to the production opportunities afforded by their natural resources and their markets. Ranch operators had to do this or "go broke." But there are yet a good many ranches in the process of failure because they are fundamentally ill adapted to their production economics and markets. And there are many ranches that are outstanding successes primarily because they have made an unusually good adaptation to their natural production opportunities. We will survey this briefly, noting some typical instances.

Owing apparently to the favorable climatic situation which has prevailed in the northern plains rather generally since 1938, a large number of the sheep ranches in this region are shifting to the use of crossbred ewes in order to attain higher lamb production. More than likely most of these ranches are newcomers in northern plains sheep production. Drouth will come again — it has recurred regularly in the past — and the sheepmen who have retained their Rambouillet ewes will stay in their business of selling feeder lambs and fine wool. They will also probably buy out the ranch properties of those who undertook to grow range-fat lambs in the plains grasslands.

This favorable climatic situation was also instrumental in inducing some of the plains cattle ranches to produce two- and three-year-old steers for the slaughter markets. But this change was induced mainly by the cattle and feed-grain price relationship that prevailed nationally during the World War II period. Most of these ranches have returned to calf and yearling marketings or soon will.

Adaptation of the ranch to the best production opportunity of the natural resource becomes more varied and complex to the west of the plains. For illustration we look at two Montana ranches cradled in the giant apron of wrinkled and folded land that sweeps eastward from the Crazy Mountains. These ranches have the same type of rangeland. One of them has a large acreage of natural meadowland and ample irrigation water; the other is rather limited in these resources.

The ranch with the plentiful meadowland and ample hay production operates on a cow breeding herd basis and markets choice feeder calves. Two-year-old steers, usually "good" grade slaughter animals, constitute the principal marketing of the other ranch. The ranch that markets calves uses the meadow regrowth for fall pasture as added green feed for some of the cows with calves and feeds ample hay (about 1¾ tons) through the winter to the cows. The other ranch winters the cows somewhat harder and uses less hay for the yearlings and steers. Both these ranches have made a good adaptation to the production opportunities of their resources.

Adaptation of the western stock ranch to its natural resources is a real and an important management feature, much more significant in ranch operations than in crop agriculture where cultivation methods and artificial means under the control of the grower enter to a greater degree into the production processes. This adaptation is not, however, a static thing. Some rather recent developments in animal breeds and types and in ranch mechanization have influenced the natural production opportunities considerably for many ranches. As an example in cattle breeds and types, the development of the "ultramodern" type of Hereford and the "discovery" of the range potentialities of the Angus breed will no doubt increase the production and profits of ranches with suitable range and pasture resources. Some ranches with arid lands, low-capacity ranges, rough ranges, and limited winter feeds will write the history of their experiment with these animals in red ink.

Though perhaps not very numerous, there are ranches throughout the West that have not made good use of their resource opportunities because they have been unwilling to make changes requiring increased investments and increased operating costs. One illustration of this is the sheep ranches that have the upland summer range and the winter feed resources for early-lamb production, but that are range-lambing and producing a fine-wool feeder lamb. Then there are some cattle ranches with land and water sufficient for expansion of irrigated pasture production; these resources have not been utilized in the past because of the lack of rangeland to balance the added feed that might be grown. Some of these ranches are going to find in the increased West Coast markets for beef a new incentive for the development of irrigated feed-crop and pasture production. Especially will this be so if more of the animals marketed can be stepped up in grade through the use of feed-crop and pasture production.

Basing Ranch Values on Earnings. The real estate markets for western ranch properties have not been a very good guide to the sound investment values for such properties. This is partly because rather generally western stock ranches did not attain organizational maturity and stability until the twenties. And since that time a severe economic depression, the Great Plains drouth, and World War II have intervened. It isn't surprising, consequently, that the concept of normal capital values for the stock ranch, based upon a reasonably good relationship between gross earnings, operating costs, and net earnings, is regarded by many as a theory if not a fiction. But the only alternative to that concept is a "boom and bust" approach to values and valuation of ranches.

Let us review briefly the changing picture of western stock ranch economy and the influence of such changes upon ranch property values. There have been three fairly distinct periods in these changes. The first period, from about 1880 to 1905, was the time of open range operations. During the second period, from 1905 to 1925, western homesteading and land settlement caused rapid and far-reaching changes. In the third period, or since about 1925, western stock ranches have been working toward stability of land ownership and tenure and of production management. But the requisites for price and cost stability have not prevailed to such a degree that the real estate markets could develop capital values for what might be considered any normal situation of gross earnings, operating costs, and operator net incomes.

The period of open range operations was inevitably temporary and unstable. It is a somewhat useless polemic to discuss now what might have prevailed in western stock ranch economy had there been a better suited program of land settlement and of western agricultural policy. The range livestock operations of the open range period were rather generally large and extensive, the profits were sometimes fabulous, and the losses were often disastrous. There was generally little thought of, or incentive for, owning land other than for a ranch headquarters. Migration to the localities best suited for winter range substituted to a large degree for winter feed crops, but the migration possibilities were rapidly eliminated after land settlement began to close off the migration routes.

The thoughts of the range operator at the close of the open range period are aptly expressed by the following quotation from a book of western verse:

> Oh, it's squeak! squeak! squeak!
> Hear 'em stretchin' of the wire!
> The Nester brand is on the land,
> I reckon I'll retire.
>
> Their house has locks on every door;
> Their land is in a crate.
> These ain't the plains of God no more
> They're only real estate.*

During the period of western homesteading, the ranch operations began to organize their land holdings to compete with the farming operations for the ownership and control of rangeland, water, and feed-crop land. This was another transitional period for most western stock ranches. Many western areas, homesteaded for crop agriculture and since returned to a stock ranch economy, have had to wash out land values and land costs that were generated by the temporary and unsuited farm economy. This is well illustrated by a comparison of the history of land costs for the

* From "The Old Cowman" in *Sun and Saddle Leather* (Boston: Badger Publishing Co., 1922) by Charles Badger Clark. Copyright 1935, 1936 by Chapman and Grimes, Inc. Reprinted by permission.

stock ranches of the Musselshell valley of central Montana and of the Big Hole basin of southwestern Montana. By 1920 the stock ranches of the Musselshell valley were rather generally attempting to carry an annual land cost for the land taxes, land leases, and the interest on land investment, of $6 to $8 per head for the year-round capacity of the ranches. This has since been readjusted to the feasible limits for stock ranches in this area — an annual land cost of around $4 per head for the number of cattle that can be well maintained. The ranches of the Big Hole basin, where the elevation and the limited growing season preclude crop agriculture, have, until the recent war period, typically had an annual land cost of $4 to $5 per head for the cattle that can be maintained.

During the third of these periods in western stock ranch economy, or since about 1925, the stock ranches have rather generally been working toward the essential management objectives of a seasonally well-balanced unit, a sustained productive capacity, and stable land ownership and tenure. If, during the next decade, there prevails a period of reasonable price stability with prices and costs in a good relationship, we shall likely see market prices of ranch properties become a fairly good expression of the normal earning values of such properties.

These normal earning values will depend upon the margin between gross earnings and the operating costs (including a remuneration to the ranch owner for the market value of his wages as a ranch operator). This margin is the "economic rent," the remuneration to the land resource as a share in the income. This income share of the land can be calculated as a residual claim on the earnings, since the land resource is a fixed and highly specialized production factor, but the livestock and equipment capital and the labor are much more fluid. They must be paid the competitive rates they can earn in other fields.

Let us follow through a set of calculations to see how this works. We will want to see, under the most probable price, income, and production cost relationships, what the annual economic rent is per unit (cow or ewe) of annual land capacity, and what this means per acre for rangelands and haylands of different capacities and yields per acre. We will also want to see how much of this economic rent residual is taken by real estate taxes and how much of it is then available to pay interest returns on the land values. All these calculations must be based upon a ranch of sufficient size to give the operator a full-time job as a worker and a manager. Let's say this is about 250 head of cattle or a range band of 1000 head of sheep, a usual lower limit. In our calculations we will assume the ranch owner receives an annual wage of $1200, since this has in the past been approximately the wage of a foreman where a ranch of this size was operated under foremanship by an absentee owner.

Referring now to Tables 9 and 12, we obtain the gross revenue figures for our calculations. The mountain valley sheep ranches, the slaughter-lamb producers, show, under medium prices, an annual gross revenue of $6.65 per head of sheep operated. This figure for the sheep ranches producing feeder lambs — usually ranches with less productive range — is $4.90. The comparable figure for the mountain valley and foothill cattle ranch is $21.95 per head of cattle. For the semidesert cattle ranch this figure is $13.50. These figures give us the high and low points for our calculations of the economic rent of lands and the corresponding values of ranch properties as based upon earnings.

We turn next to Tables 15 and 16 for the operating costs that these types of ranches will, under a medium price situation, probably have. These costs, excepting the land cost items, are the costs that have the first claim on the ranch income, since they are determined mainly by competitive economics outside the ranches. From the cost figures shown in some detail in these tables we except the leases and grazing fees and the real estate taxes (usually about one half of the total taxes paid by the ranches), because these items are a part of the economic rent residual which we seek. We have, then, for the sheep ranches, $4.30 to deduct from $6.65, and $3.10 to deduct from $4.90. For the cattle ranches we have $12.50 to deduct from $21.95, and $7.15 to deduct from $13.50.

Thus we get $2.35 and $1.80 per head for the sheep ranches, and from these figures we subtract $1.20 ($1200 ÷ 1000 ewes) for the operator's wage. This gives us an annual economic rent residual per head of sheep of $1.15 for the range-fat lamb producers, $.60 for the feeder-lamb producers. Similarly, for the cattle ranches we get $9.45 and $6.35 from which to subtract the $4.80 per-head operator wage ($1200 ÷ 250 head) . This gives our figure on the per-head annual economic rent residual amounting to $4.65 for the mountain valley and foothill cattle ranches and $1.55 for the semi-desert cattle ranches.

On the usual 5 : 1 range and feed requirement ratio for cattle and sheep these figures compare fairly closely for the two most productive types of ranches, that is, the mountain valley early-lambing sheep ranch and the mountain valley or foothill cattle ranch. This comparison is $5.75 ($1.15 × 5) with $4.65. For the low-producing ranches, the feeder-lamb producing sheep ranches and the semidesert cattle ranches, this comparison is $3 ($.60 × 5) with $1.55. The semidesert cattle ranches are the nearest to the "no rent margin" of any of the main types of western livestock ranches.

In carrying the above calculations on into per-acre land values we are going to use a maximum annual economic rent figure of $5 per animal unit (one head of cattle or five head of sheep) as a compromise between the $5.75 for the sheep ranches and the $4.65 for the cattle ranches. For the minimum figure to represent the annual economic rent of the less productive ranches and lands we will use the figure of $3 per animal unit for the sheep ranches. The $1.55 figure for the semidesert cattle ranches is considerably below this, but the plains cattle ranches using low-capacity range show a figure somewhere near the $3.

These maximum and minimum annual economic rent figures per animal unit we will tie in with five grades of land, relating the maximum annual rent per animal unit to the best grade, the minimum annual rent per animal unit to the lowest grade, and calculating in a straight line relationship between this maximum and minimum for the values of the grades between one and five. In deriving the land capital values per acre we are going to assume that the land taxes will average about 2 percent of the land values and that the interest earnings are 4 percent on the land values. That is, taxes absorb one third of the economic rent and the other two thirds is capitalized into land values at a 4 percent rate. The first of these premises checks fairly well with the facts, and the 4 percent interest rate is somewhere near the land mortgage interest rate of recent years.

In Table 1 we have the results of these calculations for the five grades of rangeland defined. Note that for the first grade the rental value of $.21 is (in round figures) the annual economic rent of $5 divided by the annual land requirement of 24 acres; that one third of this is assigned to taxes; that the other two thirds is capitalized at 4 percent to derive the land value; and that the real estate tax is about 2 percent of the capital value of the land. This table expresses the tax rate as an upper limit rather than as a standard. It seems likely that a rangeland tax rate above 2 percent on these values would have an adverse influence upon the incentive for private ownership. Present rangeland taxes are, however, frequently more than the 2 percent of the values shown in Table 1.

TABLE 1. CAPITAL VALUES, RENTAL VALUES, AND TAX COSTS
PER ACRE FOR RANGELANDS

Grades of Rangeland	Normal Capital Value per Acre	Normal Rental Value per Acre	Tax Cost per Acre (upper limit)
First (2 acres per cow month or 5 sheep months)	$3.50	$.21	$.07
Second (2½ acres per cow month or 5 sheep months)	2.75	.16	.05
Third (3½ acres per cow month or 5 sheep months)	1.75	.10	.03
Fourth (5 acres per cow month or 5 sheep months)	1.00	.06	.02
Fifth (8 acres per cow month or 5 sheep months)	.50	.03	.01

SOURCE: Mont H. Saunderson, "A Method for the Valuation of Livestock Ranch Properties and Grazing Lands," Montana Agricultural Experiment Station Mimeographed Circular No. 6, 1938.

In qualification of these standards it should be observed that the grades of land take into account only one quality factor — grazing capacity — which is, however, generally the most important one. Other factors, such as topography, water availability, and additional features that influence weight gain, calf and lamb crops, and death losses, can cause substantial departures from these value standards.

Normal value standards for ranch haylands are given in Table 2 on a basis comparable with the rangeland value standards given in Table 1. The per-acre capital values in Table 2 are based upon a land investment per animal month of capacity comparable with the investment per animal month of capacity for rangelands as in Table 1. (Animal months for hayland are calculated at 500

TABLE 2. CAPITAL VALUES, RENTAL VALUES, AND TAX COSTS
PER ACRE FOR RANCH HAYLANDS

Grades of Hayland	Normal Capital Value per Acre	Normal Rental Value per Acre	Tax Cost per Acre (upper limit)
First (2½-ton yield per acre)	$60	$3.60	$1.20
Second (2-ton yield per acre)	45	2.70	.90
Third (1½-ton yield per acre)	30	1.80	.60
Fourth (1-ton yield per acre)	20	1.20	.40

SOURCE: Mont H. Saunderson, "A Method for the Valuation of Livestock Ranch Properties and Grazing Lands."

pounds of hay per head per month for cattle and 100 pounds for sheep, plus the meadow grazing obtainable for the haylands.) The per-acre rental values of Table 2 do not include any of the costs of using the land; that is, these are rental values only. The value standards for ranch haylands are based upon yield only and may be substantially modified by differences in the quality of hay that can be produced.

The livestock prices that prevailed during the period from 1940 to 1949 make the preceding calculations on economic rents and ranchland values appear unreal, even though ranch operating costs did nearly double during this period. The more productive of the cattle ranches have received an annual gross income of more than $75 per head for cattle during this period. Many cattle ranches have been realizing as high a *net* income as may normally be expected for the *gross* income.

As a consequence some startling prices have been paid for ranch property and some extreme rental rates for rangelands. Some cattle ranch properties have sold at prices amounting to a real estate investment of $200 to $300 for each head of cattle that the ranch can carry. Rangeland rental contracts costing $1.50 per animal month for cattle have been executed for a five-year term. These figures are, of course, far beyond the normal value limits for the ranch that derives its revenue from range beef cattle production. There undoubtedly have been some influences other than ranch earning values in many of the extreme ranch sale prices recently, especially for ranches in the Rocky Mountain region. Competition for the foothill and mountain valley ranches in the more attractive sites has sometimes resulted in prices far above the economic limits of their income from livestock production.

The ranch property prices during the World War II period make the value standards set forth above seem ultraconservative, perhaps unrealistic. These values *are* somewhat on the conservative side. Our national population growth may enhance cattle prices sufficiently to lift western ranchland values materially. But the evidence on ranch earning power indicates that if we are to have stability in ranch ownership and in ranch management, we cannot get very far away from some such land value and land cost standards as those given here.

It should be re-emphasized, finally, that while the value standards presented here are based upon a good cross section of western ranch operations, many individual ranches will justifiably differ materially from these standards, because of local situations in resources, costs, markets, and other management opportunities and obstacles. The practical use of these standards is that of general guides against which the situation of the individual ranch can be measured, and the reasons for any material differences studied and analyzed.

CHAPTER II

Livestock Management Economics

You will not find an adequate treatment of range livestock husbandry in the published materials on animal husbandry. Only rather recently has livestock science recognized that the livestock husbandry problems and the livestock economics of western ranches are distinctly different from those of livestock production on the farm. Comparatively recent, too, is the increased orientation of the western breeders to the needs of the ranches for adapted breeds and types of cattle and sheep. To a considerable extent the western ranch husbandman has been on his own in the selection of animals and in fitting husbandry practices to the natural pattern of his ranch.*

In some degree, the selection of the kind, breed, and type of livestock and the livestock management practices will always remain an individual problem for each ranch, because of the local differences in resources and in the natural factors that are so important in range livestock production. The ranch operator, seeking the best adaptation of his livestock production to his natural factors and to his markets, asks himself these questions:

What kind of livestock, what breed, what type within the breed, will give the highest net income from my range forage and feed crops?

What kind of livestock management practices are best suited to the production opportunities of the ranch?

What ranch improvements will aid in the livestock management program of the ranch?

* For a treatment of livestock management and ranch management interrelationships with particular application to the low mountain and foothill cattle ranch of the Rocky Mountain region, see Wallis Huidekoper, *Modern Beef Cattle Breeding and Ranching Methods*. This monograph is published and distributed by the Montana Stockgrowers Association, Helena, Montana.

Let us consider the principles which afford a guide in answering these questions.

In the choice of cattle or sheep, or some combination of the two enterprises, the types of range and the kinds of feed crops that can be grown must be given first and primary consideration. Sheep have a natural preference for the range types with a predominance of browse and weed feed. Their use of the grasses is confined largely to the short grasses, the smaller of the bunchgrasses, and others of the finer grasses. They do, however, make some use of the midgrasses in the spring when they are green and succulent. Alfalfa hay and native hay from finer grasses make the best sheep hay. A much higher proportion of sheep than cattle are winter-ranged. Those winter ranges with range plants having considerable protein are sought by the experienced sheepmen. One such plant is the bud sage of some of the Nevada and western Utah ranges. Sheepmen refer to this plant as a "hot feed." They know it will help avoid weaknesses in the wool due to protein deficiency.

Cattle have a natural preference for the grasses. They use the finer grasses as well as the coarser ones unpalatable to sheep. Cattle make some use of coarse browse feeds such as snowberry, serviceberry, aspen, and willows, and they will also use the coarser native grass hay, grain hay, straw, and coarse roughages. Most sheep ranches have some such coarse feeds, and many of the sheep ranches profit by having a sufficient herd of beef cattle to use the coarse hay and the weathered hay that sheep would waste, the range growth that sheep will not eat.

In those of the western ranches where the type of range has undergone considerable change because of prolonged use by one or the other kind of livestock, a shift in kind of livestock is often profitable; it may, in fact, restore the natural balance in the range plant life. This management relationship between kind of livestock and use of range forage is most likely to prevail on ranges that are a mixture of grasses, weeds, and browse plants. Several years of heavy use of such a range by sheep may change the range type mainly to grass, and the range may then be economically better suited to cattle. After a period of use by cattle this range may revert to its original type, and then a moderate and well-balanced dual use to hold that balance may be the most economic and stable use. However, any such shifting in kind of livestock must depend in some degree upon the adaptability to cattle or to sheep of the winter feeds and seasonal ranges, if these are a part of the range forage and the feed-crop setup of the ranch.

Probably for a majority of western ranches there is a considerable margin of choice between cattle and sheep, even though many ranches are limited to one or the other kind of livestock by the type of range and the kind of feed crops. That there is a considerable margin of choice is indicated by the fairly regular cyclical swing occurring between cattle and sheep populations in the western states during the past fifty years. Sheep ranches have shifted to cattle when the cattle and sheep price relationship favored the net income from cattle. Then as cattle numbers reached the peak of the cycle, with sheep numbers down and sheep prices on the upturn, ranches start the shift from cattle to sheep.

This shifting between cattle and sheep in response to an interrelated price and population cycle has been a disadvantage to most ranches. Greater production stability and better incomes would usually have been realized by retaining the kind, breed, and type of livestock best suited to the ranches' natural possibilities and long-range market opportunities. Through the recent war period, as in World War I, the price and cost situation favored cattle over sheep. As a consequence there was some rapid shifting from sheep to cattle. This decline in sheep numbers was, however, in part due to the wartime labor problems of the ranches, since on an animal unit basis the labor requirement for cattle is considerably less than for sheep. The decline was greatest in the labor-scarce northwestern states — Montana, Idaho, Oregon, Washington.

Choice of Breeds and Types of Sheep. The choice of sheep breeds for the western sheep ranch lies between the fine-wool breeds and the mutton breeds, in the effort to attain the best possible combination of lamb and wool production afforded by the ranch. Fine-wool sheep breeds, of which the Rambouillet is the principal one used by western ranches, were derived mainly from the European Merino breed. They have the strong flocking instinct of the Merino, which improves their handling in range bands. They produce a heavier and finer fleece than the mutton breeds, but they do not have the fleshing qualities of the mutton breeds. Because of their ancestry, Rambouillet sheep have some tendency to revert to the fine wool, wrinkly skin, and small body of the Merino. This tendency can be offset by continuous and skillful selection in the replacements and by culling. Mutton breeds of sheep, such as the Hampshire, Suffolk, and Shropshire, require more favorable forage and feed conditions than the fine-wool breeds. The Lincoln and Cotswold are coarse-wool breeds with size and fleshing characteristics, but they are not rated as true mutton breeds.

A considerable proportion of the sheep ranches of the Rocky Mountain region, of California, and of parts of the intermountain region have a range situation affording an opportunity to produce slaughter lambs from the summer range. But they cannot do this to any extent with the straight use of a fine-wool breed — that is, for example, using Rambouillet ewes and sires. Some such ranches consequently use mutton-breed sires with Rambouillet ewes and either buy lamb replacements for ewes or, if the ranch is of sufficient size, breed a band of the ewes to Rambouillet sires to grow the replacements. Some of the ranches that can produce range-fat lambs have resorted to their own program of crossbreeding the mutton breeds and the fine-wool breeds to grow crossbred ewes for the improvement of their slaughter-lamb production. The results have not been very good either genetically or economically. Too much variability has resulted in the type of lamb and wool and in fleece quality. Some fairly recent work by the animal geneticists and by the purebred breeders appears to be solving this problem for the western growers. Some new adaptations of breeds and types that may alleviate the need for crossbreeding by growers appear to be forthcoming.

Rambouillet breeders have in recent years developed a considerably larger and smoother type of the Rambouillet, and apparently they have been successful in attaining stability of the type.. This breed and type is adapted to the plains, Rocky Mountain, and intermountain ranches that have a favorable range and feed situation but do not have a summer range suited to slaughter-lamb production. Then there are the comparatively new breeds, the Columbia, the Panama, and the Targhee — the so-called white-faced-crossbred breeds. These breeds have a somewhat coarser wool than the Rambouillet, and as larger sheep they are some-

The Rambouillet is the foundation breed for most western sheep ranches. (National Woolgrowers' Association Photo)

Columbia ewes. Western sheep ranches are increasing their use of the Columbia and other crossbred-whitefaced breeds. (National Woolgrowers' Association Photo)

60

what better lamb producers. Their adaptations are similar to the above-described type of Rambouillet, though perhaps they are somewhat better for slaughter-lamb production, under favorable range conditions, when bred to sires of the mutton breeds. There are some western ranches that use one of the true mutton breeds, both ewes and sires, for range-fat lamb production. Herding these sheep in range bands is difficult, and there is some sacrifice in wool income. Farm-flock operations attain high lamb production on good irrigated pastures with the use of mutton-breed ewes and sires.

Where sheep ranches operate under adverse range forage and feed-crop conditions, as they often must in parts of the plains regions and on the winter ranges and lower summer ranges of the intermountain region, there is merit in the small, hardy, fine-wool type of the Rambouillet breed. Here the operation necessarily aims at feeder-lamb production, realizing a good proportion of the income from wool, and a low operating cost.

Choice of Breeds and Types of Cattle. Until rather recently the western range cattle producers favored the large-framed type of Hereford. Their objective was a range-worthy animal with a good growth potential on range and roughage feeds (some growers made occasional use of Shorthorn sires to increase the growth potential). This type of animal matured rather slowly and did not make too good a young feeder animal for feedlot finishing. The large-framed Hereford remains the best choice for many western ranches, especially ranches with productive but rough foothill and mountain ranges and ranches with large amounts of coarse forage and roughage feeds to convert to beef.

In recent years the Hereford breeders have developed the so-called "modern" type of animal. This type, thick-bodied and low-set, matures early and fattens efficiently. Under favorable range and pasture conditions these are, no doubt, superior beef animals.

In recent years there has been a noteworthy trend among western growers to the use of the Angus breed. The Angus is comparable to the recently developed types of the Hereford breed in growth efficiency and early maturity, and this, with the shift to feeder-calf production, appears to account for the increasing favor in which this breed is held.

A type of the Angus breed has not as yet been developed, how-

Hampshires. Western sheep ranches capable of range-fat lamb production often use Hampshire sires. (Record Stockman Photo)

Suffolk sires are increasingly used by ranches that can produce range-fat lambs. (University of Idaho Photo)

ever, that is as well adapted to rough foothill and mountain range as the large-framed types of the Hereford breed are. The Angus is blockier and shorter legged — better adapted to the stock farm than to the ranch that uses rough range or low-capacity and poorly watered range. This breed has the advantage of being hornless and the black color makes the Angus virtually immune to eye cancer and to udder sunburn.

The Shorthorn breed has been used only to a limited extent by the western stock ranches. The reasons for this appear to have been the lack of a beef-type Shorthorn as range-worthy as the Hereford and the slower growth, later maturity, and larger mature size of the Shorthorn. However, some very recent work of Shorthorn breeders indicates that a type of Shorthorn is being developed that has a compactness and meat-producing efficiency similar to the "modern" type of the Hereford breed. Western stock ranches that have productive range of favorable topography may be making more use of the Shorthorn breed in the future.

RANGE LIVESTOCK HUSBANDRY PRACTICES

Let us re-emphasize here the influence of the many variations in natural environmental factors of ranches upon the adaptation of a management program to a ranch. These important variations among ranches and the close management interrelationships between range livestock husbandry, grazing land management, and feed-crop production and use make it especially important that livestock husbandry be thought of as an integral part of the management program of the ranch as a unit. Too frequently livestock husbandry has been viewed as *the* management program of the ranch rather than a phase of the management program. Practices followed in the management of livestock must be adapted to the natural factors and the best economy of ranches.

For example, a productive foothill cattle ranch with a rough breeding range may find it good economy to use a bull for each fifteen or twenty cows, and to incur extra cost for labor to manage the stock during the breeding season, although the less productive ranch would find these high costs uneconomic. The sheep ranch with an upland summer range where terrain and other factors make herding and range management difficult may find it

The large-framed type of the Hereford breed, favored by ranches desiring a range-worthy animal. (Record Stockman Photo)

The "modern" type of Hereford, early maturing and efficient on high-capacity ranges and pastures. (Bar 13 Ranch Photo)

good business to operate summer bands of 800 ewes rather than 1200 ewes.

Sheep Ranch Husbandry Practices. Throughout most of the West the ewe band constitutes the natural livestock management unit. Where winter feeding is practiced, three summer bands can be efficiently combined into two winter bands. Where sheep are winter-ranged, two summer bands make one winter band. The adapted size of a summer band may vary from as low as 700 ewes for the ranch using rough or timbered mountain range to 1500 ewes for the plains ranch. The size of winter bands varies from 1500 to 1800 ewes where the ewes are maintained through the winter on hay, from 2000 to 3000 where they are winter-ranged. Winter bands are made up at the start of the breeding season, usually middle to late autumn except for the California ranches and the Arizona spring-lamb producers. On ranches where several bands can be operated the ewe bands are usually made up on a basis of age classes.

Just before the breeding season, the ranches that can do so usually "flush" the ewes by placing them on good feed—crop aftermath, hay meadow regrowth, or some especially productive range—for a period before the sires are placed with the ewe bands. Though some ranches do not do so, most of the western sheep ranches have found it profitable to give considerable attention and management to the sires at breeding time.

A plan of rotating, resting, and graining the bucks is a common practice in the management of bucks during the breeding season. Some operators turn all the bucks in with the ewe bands at the start of the breeding season, and others reserve the younger and more vigorous sires for the latter part of the season. These management practices increase the efficiency in the use of sires, and by lowering the required number of sires, make it possible to use better sires without increasing the service cost. There are operators who, through good managment of the sires, do the breeding efficiently with an average of one buck for 60 ewes; other operators use one buck for 35 ewes, with the same type of sheep. The required number of sires is less for the mutton breeds than for the fine-wool breeds. The usual practice of the western sheep ranch is to use the sires through four or five seasons, but the average an-

More western ranches are now using the Angus. (Record Stockman Photo)

Until recently, western ranches have used the Shorthorn mainly for cross-breeding. (Record Stockman Photo)

Where possible, sheep ranches flush the ewes on green feed before breeding.
(Record Stockman Photo)

nual death loss of about 25 percent for the sires reduces the actual average length of use.

Where the ewe winter bands are maintained partly or entirely on hay, the hay feed requirement varies from about 200 pounds per ewe in the northern plains — here the hay feeding is intermittent and totals about sixty days — to 600 pounds for some of the mountain valley ranches. This is for the ranches that feed continuously for five months and have crossbred or other large sheep. Ranches maintaining the ewes on winter range often feed some protein concentrate range supplement during cold or snowy periods. This is fed at a rate of one-fourth to one-third pound per ewe daily. Early-lambing sheep ranches that use winter range usually bring the ewe bands in and place them on hay feed (often with some grain supplement) two to four weeks before the lambing starts.

Most sheep ranches strive for the maximum practical lamb crop, and this means avoiding unnecessary death loss of the lambs during lambing. Adequate labor, equipment, and feed are very important during the first few critical days after birth in saving a maximum number of the lambs born. Some larger ranches make up a separate band of ewes with twin lambs and place it under the care of a good herder on the best summer range. Some operators who strive for range-fat lambs do not, however, favor twins.

Experienced ranchmen value highly the services of a good herder with the ewe-and-lamb bands, to avoid death losses and get maximum gains on the lambs. Beyond good handling of the ewes and lambs, a capable herder will make a success of a good range management plan, a poor herder will defeat any such plan.

Rather generally western sheep ranches cull the ewes for age when past six years, though this depends somewhat on range and feed conditions and the breed of sheep. Teeth of sheep that winter on desert ranges wear out sooner; there is some difference between breeds in natural longevity. Ranches with a good local market to farm-flock operators for old breeding ewes cull earlier and heavier for age. Skillful culling of the ewe bands for age, condition of teeth, spoiled udders, and barrenness, and eliminating below-standard individuals often make the difference between good and ordinary ranch profits.

Cattle Ranch Husbandry Practices. Western cattle ranches of sufficient size for efficient husbandry practices usually separate the cattle for management purposes into the cow breeding herd, the heifers to be bred, and the steers (if steers are produced). Since the summer range period is the breeding period for the cattle ranches practicing a controlled breeding period (a surprising number of ranches do not), this classification of animals is made at the beginning of the summer range period.

Good range husbandmen advocate locating breeding cows on the summer range in bunches of about one hundred, with summer ranges fenced for such a practice if possible, and placing four bulls with each bunch of cows. Increasingly the cattle ranches are breeding heifers as yearlings, though some differing views prevail on this practice. Usually the sires are placed with the breeding herds in late June or early July and kept there for a period of two to four months, when they are removed and maintained separately for the remainder of the year. Good operators usually give the sires extra feed and care through the winter to condition them for the breeding season.

Bull service constitutes one of the most important operating costs of the ranch. The efficient husbandman never forgets that "the sire is half the herd" and that a cheap sire is no bargain. The period of service for range bulls is comparatively short, an average of about three years before they are marketed or traded to avoid inbreeding. Death loss in range bulls, averaging about 15 percent a year, constitutes a considerable factor in the cost of bull service.

Bull service requirements vary from one bull for each fifteen cows to one for each forty cows. Where the summer range is rough and timbered upland range with much of the feed in the parks, an average of one bull for each 15 cows, plus adequate labor to keep the bulls well distributed, may be the price of getting a good calf crop. To obtain a good calf crop from one sire per 40 cows the breeding range must be smooth and of good grazing capacity, comparable to fenced pastures. As a general standard, the plains ranches now use one sire for each 25 to 30 females, the foothill and mountain valley ranches one sire for each 20 to 25 females.

Some of the Rocky Mountain and intermountain cattle ranches with rough summer ranges are developing range pastures on the more favorable breeding sites, where the sires and the females are held through the first month of the breeding season. Some cattle ranches that have a high bull requirement and cost are able to reduce the cost by maintaining a herd of purebred cows and a herd bull to produce their own sires. But this is limited to the ranches sufficiently large to justify such an enterprise. Perhaps for some ranches the practice of artificial insemination will in time reduce the bull service cost considerably, although this hardly seems likely for the big majority of western cattle ranches.

Where cattle are ranged the year round on large areas of low-grade open range or on brushy year-round ranges such as some in western Texas and in the Southwest, there are some difficult obstacles to achieving a definite breeding season. Many such ranches make no attempt to do so, to their increasing disadvantage because of the shift of markets to young feeder animals of good quality and uniform age.

At the close of the summer range season those of the ranches that must do winter feeding usually make some classification of the cattle for grazing, feeding, or some combination of the two. Cows with calves to be marketed or yearlings and steers to be marketed may be put into the hay meadow regrowth to get some additional gain and condition on these market animals. With the

High-protein concentrates make a good range supplement for · young animals. Mature animals can use the less costly carbohydrate supplements. (Charles Belden Photo)

approach of the winter feeding period the bred cows and heifers are moved to the winter pastures adjacent to the hay meadows or onto the meadows. Where possible, it is a usual practice for cattle ranches to save some winter range adjacent or close to the hay supplies, and to combine winter grazing with limited hay feeding. Such limited feed of hay may be 8 to 10 pounds per head daily, or approximately half of a full hay feed.

For many cattle ranches having winter range possibilities and somewhat limited hay supplies it is economy and good management in the winter maintenance of cattle to classify them for range maintenance and for winter feeding and to work and reclassify them through the season for winter grazing, combined grazing and feeding, and full hay feeding. An experienced and competent range husbandman can judge when an animal is going to need feed where an inexperienced operator would not be able to until such an animal slips to the point where it must be taken into the "hospital bunch" for extra feed and care through the winter season. Usually the breeding herd is the first to be put onto full winter feed. Bred cows and heifers are as a rule fed to make some gain, though not any very considerable one, through the winter. Some ranches winter-range the dry cows, yearlings, and steers with no additional feed other than cottonseed cake or some other supplement during cold or snowy periods. The feeding rate for this is usually about one pound per head for each day it is fed.

Hay feeding requirements for the cattle ranches of the northern plains, over a series of years, average about half a ton per head. Foothill and lower mountain valley cattle ranches of the Rocky Mountain region must feed three to five months and they have a hay feeding requirement of a ton to a ton and a half per head.

High mountain valley cattle ranches that use some of the hay to fatten mature animals classify the cattle into a ranch herd and a beef herd at the start of winter feeding. Over a period of four to six months the ranch herd is given a daily feed of hay somewhere between 15 and 20 pounds per head; the beef herd is fed 20 to 25 pounds per head daily.

In recent years western cattle ranches have of necessity emphasized more and more those husbandry practices that would increase the calf crop. With the shift in the marketings to younger animals it is no longer financially possible to "run a steer outfit with cows." In the past some ranches with calf crops averaging as low as 40 to 50 percent have almost seemed to be doing that. Culling the cows for age (at ten years or less), eliminating shy breeders and poor individuals, and separate pasture handling of heifers at calving—these are illustrations of practices that are receiving added emphasis.

RANCH IMPROVEMENTS FOR LIVESTOCK MANAGEMENT

Range Fences. A plan of fencing for the rangelands of a ranch should aim at the dual objective of livestock husbandry and rangeland management. In their fencing plan for the summer range, many cattle ranches achieve the range management objective of getting uniform use of the range and the livestock management objective of distributing the sires among the females by fencing the range into blocks that will handle about one hundred cows through the season.

Where yearlings or older animals are marketed the fencing plan may include separating such animals from breeding herds on summer range. Such separation would also apply to yearling heifers where heifers are not bred until two years of age. Even when bred as yearlings, some operators segregate them from the cow herds. Rocky Mountain and intermountain cattle ranches that still market considerable numbers of older beef animals sometimes fence a block of good wheatgrass range for use as a fall finishing pasture for the steers and dry cows.

Though the electric fence is yet largely experimental as an aid in the management of livestock on rangelands, it does appear to have some possibilities. Sometimes a temporary fence is needed to restrict the movement of cattle within or off a certain limited

With natural shelter, feed, and easy access to water, cattle winter well, even though wintry winds roar in the tops of the cottonwoods.
(George Sim Ranch Photo)

area of a range. One such possibility is the necessity of keeping cattle out of a poison plant area during the season of danger, which may be only a part of the range season.* Another possible use is to confine cattle for a time to a part of a range where the range forage is highly seasonal and will be lost unless it is used within a limited period of time. An illustration is the tabosa grass that grows abundantly on certain "patches" or areas (evidently on certain soil types) of the rangelands of some western Texas and southern New Mexico ranches. This grass makes good feed during its season of growth but becomes nearly worthless soon after growth stops.

Livestock Shelter. Western cattle ranches make very little use of artificial shelter for range cattle. Brush, trees, and cut-banks along the stream bottoms, close to the hay meadows and to open running water, usually constitute the shelter. This combination makes an ideal shelter, even though winter winds roar in the tops

* See Table 9 in *A Pasture Handbook,* U.S. Department of Agriculture Miscellaneous Publication No. 194, April 1934, pp. 60–61.

of the cottonwoods. Range cattle can withstand extremely low temperatures for a time if they have protection from wind, adequate feed, and easy access to running water. Most cattle ranches use the barns, sheds, and artificial windbreaks primarily for the "hospital" animals and the work stock.

Range sheep also do not require much shelter in northern climates other than naturally sheltered locations for winter feeding grounds. Sometimes artificial windbreaks (board fence panels) are constructed to make a location more sheltered. Lambing sheds are, however, generally used for early lambing in northern climates. A well-constructed lambing shed with panels for individual inside ewe and lamb pens or "jugs," with hay bunks, feed troughs, water supply, and the like requires a sizable investment. Then in addition there are the outside pens for handling the ewes in small lots for a time after they are moved out of the shed and before they are recombined into a band. Where a shed can be used through a period of four to six weeks and can serve a number of ewes equal to three summer bands, the investment is used a good deal more efficiently.

Lambing sheds are often the cause of a livestock sanitation problem. This can be met to a large extent by arrangements to admit direct sunlight into the shed on warm days during use and for a time following the season's use. Sometimes the upper parts of the south exposure of a shed are covered with removable canvas. Another similar arrangement is detachable wooden roof panels.

Sheep ranches that lamb on the range in April and May have a hazard of late-season cold storms in the northern plains in the Rocky Mountain region. Ranches with such a hazard have in one season avoided lamb losses sufficient to pay the cost of an adequate supply of the small tepee tents used to shelter a ewe and its lamb on the range. As a rule the ewe and lamb are kept in this shelter for just a few hours — long enough to dry the lamb and suckle it.

SOME LIVESTOCK MANAGEMENT STANDARDS

Standards for the Cattle Ranch. The results of the husbandry methods and practices of the cattle ranch are measured mainly in terms of the following: (1) size of the calf crop, (2) weights of animals marketed, (3) market class and grade of animals sold, (4) age of animals marketed, and (5) death losses.

Lambing sheds such as these require a considerable investment.
(U.S. Forest Service Photo)

Canvas panels used as roof covers for lambing sheds can be removed to let in sunlight. (Bureau of Land Management Photo)

Shed lambing aims at high lamb production.
(Bureau of Land Management Photo)

A calf crop of 85 head weaned for each 100 females in the breed-ing herds during the breeding season is an attainable standard. It is seldom possible to attain more than a 90 percent calf crop under range conditions. The recent average calf crop for western ranches has been around 70 percent. If a ranch is to make weaner calves its principal marketing, it must have an average calf crop of 75 to 80 percent. Otherwise, the feed and other costs that go into carrying dry cows must be realized through the sale of the dry, fat cows in the fall, and all the heifer calves are required for replacement. The total tonnage of beef marketed may not be very much less through the sale of a considerable number of dry cows, but the average price received for the total weight of the animals marketed will be less. Where the management problems of the ranch make it impossible to attain good calf crops, the best alternative is a policy of mixed marketings — yearlings, steers, and cows.

The weight of calves at weaning, usually six or seven months of age, varies regionally and locally from 325 to 450 pounds. The weights of long-yearling animals, about eighteen months of age, vary between 600 and 800 pounds, and long two-year-old steers between 800 and 1000 pounds. The variation in range cow weights is usually between 800 and 1200 pounds with 1000 pounds a fair average. A good average calf weight is about 400 pounds, although calf weights up to 475 pounds are possible under favorable feed conditions. A fair average yearling weight is 650 to 700 pounds, but 800 pounds is an attainable standard (generally with the use of some extra feed to the calf through the winter).

The low point in long-yearling weights occurs on some of the yearlong ranges of Texas and New Mexico, where brushy range impedes both range and livestock management, and both the quality and market weights are inferior. (It is difficult to be at all certain about the age of the animals that come off these ranges.) A weight of around 900 pounds is a fair average for a long two-year-old steer from the western stock ranch, but 950 to 1000 pounds is attainable on the foothill ranches of the Rocky Mountain region.

Range and livestock management are the important determinants of the weights of the animals marketed. This is demonstrated by adjacent ranches operating under similar conditions in southern New Mexico — one producing an average long-yearling weight of 550 pounds, the other 800 pounds. It is not possible to say just what part of this difference is attributable to livestock management and what part to range management. The high-producing

Good calf crops are now a requirement for profitable ranch operation.
(Oliver Brothers Ranch Photo)

ranch was stocking its range at about 60 percent of the intensity of the low producer, but it was producing a higher market turnoff of beef per section of land and approximately twice as large a net income.

Market classes and grades, determined principally by the breeding, age, and condition of the animals when marketed, reflect in large measure the success of the livestock management program of the ranch. Most of the western cattle ranches can produce young animals (calves and yearlings) in the feeder class that will grade "good" and "choice." Many ranches are not yet realizing this possibility, generally because they have a problem of breed improvement. Ranches of the Rocky Mountain and intermountain regions that market older animals, two-year-old steers and dry cows, are often able to produce animals selling in the slaughter class. This varies some with the season. When these animals sell as slaughter animals they should sell in the "good" grade. Some ranches with an excellent foothill or mountain summer range are able to market steers and fat, dry cows from the range as "choice" grade slaughter animals. This requires superior management of both livestock and range. Many more ranches can attain this level—those with a beaten-out summer range cannot.

The age at which cattle are marketed influences the production of the ranch, particularly as there may be a difference in the efficiency of weight gain between different age classes. This varies regionally and locally with the type of range and feed used by the ranch. The range gain is most efficient for yearling animals in the central and southern plains where the range is primarily a short-grass type. The short grasses of the plains are better for producing weight gains than for milk flow from the cows after the grasses cure in midsummer. As a result, the yearling—in this period of most rapid natural growth—will produce more tonnage of beef from this type of range than either the cow and calf or the two-year-old animal. In contrast with this, the foothill and lower mountain ranges of the Rocky Mountain region that produce some green feed throughout the summer will produce the most beef through the cow and calf (if a good calf crop is attainable).

Annual marketings of the cattle ranch have, in recent years, equaled 30 to 50 percent of the number of cattle operated. This averages close to 50 percent for the Great Plains ranches, which

Tepee tents help save the lambs in the plains.
(Montana Woolgrowers' Association Photo)

market a high proportion of young feeder animals, and drops to about 30 percent for the intermountain region because of lower calf crops and the marketing of older animals to California and other western markets.

Death loss sometimes becomes a very important measure of the competence of the livestock husbandry of the western cattle ranch. Even the average annual death loss of 2 percent, usually considered a safe margin, necessarily has some effect on the production of the ranch. Any figure much above 2 percent is likely to be an indication of trouble. It is seldom possible to avoid some death loss, however, even with the best of livestock management. The losses of the cattle operations on low-grade open range usually average 5 percent and may go considerably higher. The most common causes of death loss are poison plants, calving, bloat, accident, and starvation.

Death loss of cows is influenced to some degree by the age to which the cows are carried. A cow's usefulness under range conditions is generally over at about ten years of age, and some experimental material recently released indicates that a cow's efficiency as a calf producer decreases rapidly after eight years of age.* The number of yearling heifers needed, as an annual average, to re-

* B. Knapp, A. L. Baker, J. R. Quesenberry, and R. T. Clark, *Growth and Production Factors in Range Cattle*, Montana Agricultural Experiment Station Bulletin No. 400, 1942.

place the marketing of aged, cull, and dry cows, and the death loss, has been about one fourth of the number of the breeding animals. *Standards for the Sheep Ranch.* Primarily, these are the standards by which the husbandry program of the sheep ranch can be measured: (1) the size of the lamb crop in relation to the number of ewes, (2) weight and quality of fleeces, (3) weights and market classes of lambs, and (4) death loss of ewes.

Fifteen years ago the average western lamb crop was about 75 percent (the ratio of the number of lambs raised to the number of ewes in the bands during the breeding season). The average in recent years has been about 85 percent. This increase is due primarily to improved management during breeding and at lambing time, principally the latter. Losses of lambs during lambing has been a frequent cause of low lamb crops. The average lamb crop produced by western ranch operations varies for the most part between 60 and 100 percent. Ninety percent is an attainable standard as an average over a series of years for most operations. The sheep ranch with an average lamb crop as low as 60 percent is primarily a wool-growing operation. Breeding-ewe replacements for a normal death loss of ewes and for the aged and cull ewe marketing takes all the ewe lambs from a 60 percent lamb crop.

According to the production and price relationships that have prevailed during the past twenty-five years, about 55 percent of the income from sales has been from lamb and 45 percent from wool, as a western average. As breeds and types of ewes have been developed to increase lamb production without a sacrifice in fleece weight and quality, or with some improvement in wool production, the trend in recent years has been toward a higher proportion of the income from lamb. The sheep ranches of western Colorado, southern Idaho, and California that market a high proportion of their lambs as slaughter animals now normally realize 60 to 70 percent of their revenue from the sale of lambs. In contrast with this the income of the western Texas and eastern New Mexico sheep ranches is, over a period of years, usually about equally divided between lamb sales and wool sales.

Fleece weights of western range sheep usually vary between 7½ and 9½ pounds. These are grease weights; the clean-weight figures are more significant. The variation in shrink (from a grease weight to a clean weight) runs from 45 to 65 percent. Thus, it is possible for a wool clip with an average grease weight of 8 pounds to yield considerably more wool than another clip with an average grease weight of 9 pounds. The quality of fleece depends primarily upon the length, strength, and uniformity in diameter of the fibers. Many western sheep ranches have a real problem of breeding and selection for wool improvement.

Lambs marketed from ranches of the northern plains region average between 60 and 65 pounds; those from the southern plains average around 60 pounds. Ranches of the Rocky Mountain region that use the mutton breeds market lambs with an average weight between 85 and 95 pounds. Ranches of the Rocky Mountain and intermountain regions that use the fine-wool breeds and have a good summer range produce a lamb averaging about 75 pounds. The California spring-lamb producers have an average lamb weight of about 80 pounds.

Market classification of lambs as feeders or slaughter animals depends on the weight and condition of the lambs as they come off the range. When the average weight of lambs from a ewe band is 75 pounds, usually about a third of the lambs will have sufficient weight and "bloom" to go into the slaughter class. When the lambs are a mutton breed or crossbred, most of them will go into the slaughter class if they average 85 to 90 pounds off the range. The grower who can produce a slaughter-class lamb from the range has a natural economic advantage in lamb production. He is the producer of a finished product.

In considering death losses as a measure of the husbandry of a sheep ranch it should be noted that death losses for range sheep normally average considerably higher than for cattle and that much more of the death loss for ewes than for cattle is due to age. Usually the aged breeding cow is much more acceptable as a meat animal than the aged ewe is. Conversely, the aged ewe does not lose efficiency as a lamb producer to the extent that the aged cow does as a calf producer. As a consequence the range cow is usually marketed before she reaches the limit of a normal life period.

Over a series of years cows account for about 30 percent of the marketings of the cattle ranches, ewes 5 to 10 percent of the marketings of the sheep ranches. An average annual death loss of 6 percent for ewes constitutes an acceptable standard; more than 10 percent is high. The sheep ranch with a reasonable ewe death loss

Livestock management facilities should be located so that the least possible movement and disturbance of the livestock is necessary. (Montana Stockgrowers' Association Photo)

and the usual culling practices for age, and the like, must annually retain or buy ewe lamb replacements equal to one third of the breeding ewes in order to maintain the number and the normal age composition of the ewe bands. To say it in another way, on ranches where a range sheep population is being maintained, the ewe lambs held for breeding-ewe replacements constitute approximately 25 percent of the total range sheep population.

LABOR REQUIREMENTS FOR RANGE LIVESTOCK MANAGEMENT

The labor requirement of the ranches that have to do considerable winter feeding and have to produce their own feed averages about twice that of the ranches operating under year-round grazing. Cattle ranches of the Rocky Mountain region average a man-year of work time for each 100 to 150 head of cattle maintained on the ranch through the year. This includes all labor — the labor for livestock management, for feed production and feeding, and for ranch maintenance. The cattle ranches of the southern plains and the southwestern region average a man-year of work time for each 200 to 300 head of cattle. The northern plains cattle ranches handle 125 to 200 head of cattle per man-year of work time.

Sheep ranches of the Rocky Mountain region handle an average of 400 ewes per man-year of work time. This includes all the labor that goes into the ranch operation. The ranches that practice early lambing and produce all or most of their feed have a labor requirement of one man-year of work time for each 300 ewes operated. Sheep operations of the intermountain region that range all year long and winter-range with large bands of sheep (up to 3000 ewes) have an average labor requirement of one man-year of labor time for 750 ewes.

Essentially, three kinds of work make up the labor requirement of the western stock ranch: (1) the labor that goes directly into the management of livestock — herding, lambing, camp tending, riding, feeding, and the like; (2) the labor for feed-crop production, such as irrigating and haying; (3) the labor for ranch maintenance and development — fencing, water development, and repair of buildings and equipment.

All these types of work may be performed by the same person or persons on the small ranch. The larger ranch has considerable opportunity for specialization in the use of labor. Some of these types of work are highly seasonal and require a good deal of extra seasonal labor. This is particularly true for the work of feed-crop production and for handling sheep at lambing time where early lambing is practiced. The problem of obtaining and managing seasonal labor on the large ranch can be solved to a large degree by mechanization of the crop production, but this cannot be done to any extent in the management of livestock. This is one of the real management problems of the future for the large stock ranch, especially for the sheep ranches that are organized for early-lamb production.

CHAPTER III

Grazing Land Use

Range management science recognizes four main principles for the management of rangelands: (1) use a range with the adapted kind of livestock, (2) graze the range during the correct season, (3) limit the use to the capacity of the range, and (4) distribute the use well over the range. There will nearly always be some practical limitations in applying these principles to the management of the ranch. Good management, however, holds these limitations to the minimum.

Adapted Kind of Livestock. The choice between cattle and sheep (or some combination of the two) for the best use of a range is likely to involve some compromise between ecologic and economic considerations — between the natural features of the range and the management requirements of the ranch as a unit. The natural features influencing the choice of the kind and class of livestock are, principally, the kinds of range forage, the terrain and other factors affecting the comparative ease of handling cattle or sheep on a range, and the availability and distribution of water for the livestock.

Range plant growth or range type may limit the use of a range definitely to one kind of livestock. A mountain range producing usable green weed feed that is available only during the growing season is a sheep range. Rangeland where the plant growth is primarily the midgrasses and tall grasses is definitely cattle range. Between these two extremes are many combinations that may be better suited to one or the other kind of animal. The relative grazing capacity of ranges where the feed might be used for either cattle or sheep can be expressed as the "conversion ratio" for

84

changing the use from one kind of animal to the other. Where this conversion ratio is three or fewer ewes to one head of cattle (over a year old), the economic use of the range is likely to be cattle rather than sheep. Where the ratio is six or more ewes to one head of cattle, sheep are likely to be the most economic users of the range.

Where the plant growth of a range is well balanced between cattle and sheep feed, a plan of dual use may be feasible in the management program of the ranch. For illustration, many of the mountain valley sheep ranches of the Rocky Mountain region have some lower foothill and benchland range with a good growth of the coarser bunchgrasses, such as bluebunch wheatgrass, which are not well used by the sheep. Dual use has the possibility both of improving the balance in the forage production and use on a range, and of utilizing a range with double intensity in the competition of cattle and sheep for feed on a heavily stocked range. When the science of range management is more generally understood and applied there can safely be more dual use of the ranges that offer such a possibility. For some types of range, dual use will give better management than any attempt to approach full use of the range forage by one kind of livestock.

Certain features of the range terrain — especially the topography and the elevation — may influence the choice of sheep or cattle for a particular range. Extremely rough topography with the usable range forage in patches or pockets may make the handling of a band of sheep impractical. Steeply sloping mountain lands can sometimes be better used by a band of sheep because there are no practical means for distributing cattle use on such lands. In the use of some of the very high mountain lands both cattle and sheep suffer respiratory troubles ("Lunger" disease in sheep, "Brisket" disease in cattle). High elevation is, however, more of a limitation for cattle than for sheep.

Sometimes the availability and the kind of water determine whether a range is used by cattle or by sheep. Some desert ranges have no water except that obtainable during the winter and spring months from the fringes of the snow on the higher elevations. This is an important factor in limiting the use of such range to sheep. Sometimes the snow fails as a water source and water must be hauled to the sheep. There are some upland ranges where range water is practically unobtainable, yet they can be grazed for a period of four to six weeks with sheep. The sheep obtain their necessary water from the succulent green vegetation and from the morning dew on the plants.

Usually these natural factors determine the kind of livestock which can make the best use of a range. Sometimes, though, economic considerations cause the use of a range to differ from the natural adaptation. For one illustration, some of the high mountain lands of western Colorado are naturally better suited to sheep grazing than to cattle. These lands are, however, used for cattle grazing because the ranch properties are better suited for the winter maintenance of cattle. Another and similar illustration is the use of some of the low and dry mountain lands of Nevada as summer sheep range. These lands are not naturally well suited to such use, but the lack of lands to grow winter feed for cattle and the availability of sheep winter range favor use by sheep.

Correct Seasonal Use. Seasonal growth characteristics of the range forage plants largely determine the best season or seasons for the use of a range. There may be other and secondary reasons: the availability of water, for example, or the need for natural shelter on a winter range.

On the distinctly seasonal ranges of the Rocky Mountain and intermountain regions the season of range plant growth changes as the elevation causes differences in temperature, moisture, and snow cover. (The elevation factor may be locally modified by differences in slope and aspect.) Thus it is possible to determine for each of the natural vegetative zones or range types (for example, the Alpine) the usual or probable date when the growth of a range should be ready for use. Use of a range before that time reduces the season's forage growth and weakens the plants.

In contrast with these distinctly seasonal ranges to the west of the plains, the same grazing lands are used through the entire range season in the plains. In the southern plains this season is yearlong. Thus the plains rangelands must often be used during their early season of growth. What is the "proper season" of use for such lands? They do not need to be used every year during the early season of plant growth. Deferred and rotation management of rangelands applies especially to the plains ranges; it is in fact the only way that seasonal use can be applied to such lands. The adaptation of such a program of range management will vary materially between ranches. The principle of the program is, however, the same for all plains rangelands. Use of part of the range is deferred each year until well into the growing season and such deferment comes in rotation to the same part of the range each second, third, or fourth year depending on how many different parts or separations of range are used in the plan.

For an illustration of one of the varied applications of a plan of deferred and rotation management to avoid constant unseasonal use of plains rangelands, we look at a northern plains ranch east of the Big Horns in Wyoming. On one part of the rangelands of this ranch the western wheatgrass predominates, on another part the blue grama constitutes most of the forage. Each of these two kinds of range has been fenced into two subdivisions. Alternately, use of one section of the wheatgrass range — the early range — is deferred until about mid-June, and likewise use of one section of the grama range is deferred until the latter part of July. Deferred and rotation management of season-long or yearlong ranges such as those of the plains avoids such continued use of the early-growing plants as would deplete them and so unbalance the capacity of the range through the season.

Use of the interdependent seasonal ranges to the west of the plains within their adapted seasons usually means waiting until the range growth is sufficiently advanced. Where two or more interdependent seasonal ranges are used within their adapted season the grazing capacity balance between such ranges can usually be maintained or improved. But an unseasonal use of the spring range may throw such a range out of balance with the summer range, and as a result this summer range is likely to be used too early.

This has in fact occurred on a good many ranches. A natural lack of spring range or a desire to save hay by early use of spring range starts the trouble. Where this problem has developed, several measures may aid in restoring the natural capacities and balance between these seasonal ranges. One such corrective measure is the seeding of early-growing grasses on the best adapted sites of the spring range. Another is shifting part of the early spring grazing from spring range to hay meadows, when this can be done without too much detriment to the hay production.* In some localities rye pasturage produces early spring grazing successfully. Sometimes a study of the possibilities of a ranch for more productive types and varieties of winter feed crops may furnish a solution for the problem of unseasonally early use of the spring range.

While a definite plan for the best probable dates and period of use of seasonal rangelands should be a part of the management program of any ranch that uses such ranges, there will be practical and desirable departures from the plan. Yearly variations in the season of growth of range plants can make a considerable difference in the time when a seasonal range is ready for use and the length of the season the range can be used. A poor season for the production of feed crops may necessitate an earlier use of the range the following spring than would be desirable as a usual practice. It is possible, however, to know fairly definitely when the plant growth of a seasonal range is ready for use, and it is desirable to start use as near to that time as practically possible.

Use within Grazing Capacity. The grazing capacity of a range is best expressed as an annual average number of animal months. Where the acreage is known, expressing the grazing capacity of a range in terms of the acreage requirement per animal month gives a means of comparing the capacity and productivity of different ranges. Up to the present time the principal method for determining the grazing capacity of a range has been to observe the response of the range plants and soils to the rate of use that has been or is being practiced.

This method for determining the capacity of a range has some limitations, but it usually gives a fair approximation. In the application of this method, a field survey is made to get an inventory of the range plants (composition, density, vigor) and the soils. If the history of the use of the range is known, the results of the survey give some measure of the extent to which the past use has been above, below, or close to the capacity of the range. If the past use is unknown, it is then necessary to observe for a time in the future the response of the range plants and soils to a rate of use

* George Stewart and Ira Clark. "The Effect of Prolonged Spring Grazing on the Yield and Quality of Forage from Wild-Hay Meadows," *Journal of the American Society of Agronomy*, vol. 36, no. 3 (March 1944).

based upon the general standards developed for rangelands of a similar type.* Meanwhile several years of observation may be needed to determine from the response of the range plants and soils whether that rate of use is correct, too heavy, or unnecessarily light. Range science is developing what appears to be a simpler and more direct method for determining rangeland grazing capacity and this may in time be perfected. This method involves measuring (by sampling) the usable, digestible nutrients produced by a range and comparing this with the known requirements for animal nutrition. But even this will have limitations. For example, two ranges with an equal production of nutrients may have a marked difference in stability of the range soils.

Rather generally the ranch operators have in the past judged year by year the capacity of their ranges by observing the gains and conditions of the animals, not by observing any changes and trends in the range. But it is possible, on some types of range, for a change in the animal production to lag several years behind a downward trend in the condition of the grazing lands. There are several situations that may make this possible. For example, a decline or an uptrend in the vigor of range plants may proceed slowly through several seasons. Or the animal response to a trend in the condition of a seasonal range may be obscured, for a time, by a different effect on a complementary seasonal range or by an increased use of feeds.

Even with the best of information on the grazing capacity of a range there will always be a need for good observation of the trends that may be occurring in the composition, density, and vigor of the range plants and in the erosion condition of the soils. When an uptrend or a downtrend has been in progress over a series of years, it may be difficult to know what changes have occurred in the type of plant growth. This is especially true where the trend in the condition of a range has been downward over a

* For illustration, a general standard has been developed that the sagebrush-grass ranges of the intermountain region, when used as spring-fall sheep ranges, will require ¾ of an acre per sheep month if in good condition, ½ to 2 acres if in fair condition, 3 to 4 acres if in poor condition. Good condition for this type of range is defined as the situation when palatable grasses and weeds constitute approximately half of the usable forage; poor condition, when the sagebrush constitutes three fourths or more of the usable forage. ("Annual Report for 1940." Intermountain Forest and Range Experiment Station, U.S. Forest Service, Ogden, Utah.)

period of time and there have been important shifts in the type of range plant growth.

However, an experienced observer of range types and of range plants and soils can usually read the story of the original native type of plant growth and can tell what management measures may restore the former plant associations and the productivity of a depleted range. For example, a trained observer knows when the Indian ricegrass and the fescue grasses are a natural association of the piñon-juniper range type, even though such grasses may have largely disappeared because of unseasonal use or overuse, and he can judge the management measures that may restore these grasses.

One aid for determining the trend in the condition of a range is the use of range check plots where changes in the plants are measured annually against the recorded readings of the first year of observation. The trend that such measurements may reveal indicates whether the level of use should be reduced or whether it can be increased. Another aid for determining trends in the condition of a range is the use of fenced, ungrazed plots for comparison with the grazed land.

Practical ranch management considerations in some years cause a departure from the grazing capacity standards set up for the use of a range. Some ranges naturally vary considerably from one year to the next in the growth of the range forage. Not many ranches can easily vary the livestock numbers from year to year to meet fluctuations in range forage growth. However, certain use limits should be observed in the annual departures that may be made from the average annual grazing capacity of a range.

Range management science has developed certain guides to such departures in so-called utilization standards — estimates of the amount of annual range forage growth that may safely be taken and how much should be left. These standards are expressed in terms of both the percentage of ungrazed plants and the percentage of the season's volume of growth that is left ungrazed.

The standards and limits of use for these measurements vary with different types of range. For a bunchgrass range the desirable use standard generally is to leave about one fourth of the plants and one half of the season's growth volume unused at the close of the grazing season. For a range of mixed type, where the animals using it have a preference for certain species of plants, it may be desirable to emphasize the measurement of the use of these plants as "key species" in judging whether the season's use is within desirable limits.

Any average grazing capacity figure has limitations as a management concept if there is considerable fluctuation in plant growth due to climatic variation. Where such fluctuation is common, as on the grasslands of the plains, more emphasis should be placed on observing the part of the annual forage crop used and the part left unused, less on average grazing capacity standards. Where the annual variations in plant growth are not extreme, an average grazing capacity figure is a more usable management concept.

In closing the subject of grazing land capacities, it may be said that management has too often been passive, accepting as final the production established by nature through plant species and climatic and site factors. To the extent that the economics of each locality justify, it may be assumed that in the future more range-lands will be treated with various cultural methods. These methods may be seeding, flood and ditch irrigation, contour furrowing, the introduction of various seasonal forage species, and other practices which should lift grazing land capacities somewhat above natural limits. But the use of any such cultural methods should be based on an analysis of the needs of the ranch for improvement of the balance in the seasonal feed capacities that make up the year's feed supply.

Grazing Use Distribution. Distribution of use by livestock evenly over a range is attained principally by the plan of range fences and water development, the location of salt, and control of the movement of animals through herding and riding.

Use of fences for the distribution of grazing use applies particularly to cattle range, since the movement of range cattle is not controlled by riding to the extent that the movement of the range band of sheep is controlled through herding. Through natural preference cattle would leave some parts of a range lightly used and overuse other areas. One illustration of this is in the use of north and south slopes of foothill and mountain range. The north slopes, except where timbered, usually produce more feed, but they are likely to be lightly used unless the cattle are held on them for

Range fences can be planned to serve the needs of both range management and livestock management. (U.S. Forest Service Photo)

part of the season. Another illustration is the natural tendency of cattle to overuse the range near water and shade on rough upland ranges, while ample range forage goes unused on the adjacent accessible slopes. Where fencing is not practical, as it may not be on rough upland ranges, placing salt at a strategic location between the watering places may help to even the use between the easily accessible and the less accessible feed.

Western cattle ranch operators have often planned the fencing program of their ranges primarily for livestock management and only incidentally for range management. The following quotation, taken from the report of a management analysis of a large Rocky Mountain foothill cattle ranch by a competent ranch manager, gives a good illustration of this.

There are a number of things that can be done to make this ranch more productive and I am taking the liberty of pointing out the most obvious things that should be done and can be done at a very little expense.

First of all there must be brought about a change in attitude toward the use of the range resources of this ranch. Three-fourths of all the feed resources are tied up in the range and 85 to 90 percent of all the beef gains are made in harvesting the grass, yet the knowledge displayed by the present employees as to the importance of this phase of the operation is indeed very meager, to say the least. I doubt if much real thought or leadership has been given to this all-important matter by the owners. It may not be any particular mark of bad management, rather a lack of management and probably lack of knowledge of what the problem is and how to solve it. As I rode over these ranges I was struck by spotted utilization of the various pastures. Some were grazed very close while others had not been used at all, or if so, very lightly.

Fences are used almost entirely to facilitate the handling of stock rather than for controlling the use of the range according to seasons, adaptability, topography, etc. To illustrate, the main streams flow east, so that there are large areas of southern slopes along these streams, but the fencing cuts across the streams so the south slopes and north slopes are in the same pastures.

Now it is important on this ranch to reserve all south slopes for spring, fall and winter use, but no such provisions are made. Where south slopes are reserved at all, the enclosures include north slopes as well, and where the ranges are used in the winter there is little use of the north slopes because they are often covered with snow and are always colder than the south slopes which are preferred by cattle. Considerable feed is thus wasted on the north slopes. It is always difficult to get full and proper use of north slope ranges, yet they produce 50 to 60 percent more volume of feed in foothill country than do south slopes.

In this northern country where winter feeding periods are long at best, every reasonable effort should be made to take advantage of those slopes of range areas that can be used early in spring or late in the fall. It is much cheaper to graze livestock than to feed hay. A careful study should be made of the entire range to work out the best possible seasonal use. I am sure that such a study would greatly increase the spring-fall use of the range lands and reduce the amount of winter feeding. If this spring and fall use could be increased 30 days on 50 percent of the cattle, it would mean a saving of 250 to 300 tons of hay, an item well worth while.

Water is sometimes the limiting factor in getting good distribution of use on a range. For good range management all parts of a range should be accessible within one and a half miles from water when the topography of the range is reasonably favorable — level to rolling. The limit should be about one mile for lands of rough topography. A range that is not adequately provided with livestock water cannot be well managed; parts of such a range will be overused while other parts will remain underused or unused.

Mountain and foothill rangelands are usually well watered, with the facilities in the form of developed springs and seeps piped into troughs and tanks. Frequently, the average distance to water on these lands is not over a half mile. This is ideal and desirable on high capacity ranges where the water can be developed at low cost.

Range water developments in the plains are usually earth reservoirs to store runoff, and wells powered by wind and small gasoline engines and having storage facilities. The use of wells predominates in the southern plains region, often with costly equipment for large storage and for piping from storage tanks to small tanks at some distance from the wells. Investments in water facilities by the cattle ranches of western Texas, New Mexico, and Arizona run as high as $30 to $40 per animal year of land capacity. Range water development costs are often as important in the investment in lands and improvements by the southern ranch operations as the development costs of hay and feed-crop lands for the northern operations.

The availability of natural water is important in determining the management and the seasonal use of many of the low-capacity grazing lands of the intermountain region. Costly well and reservoir water service is out of the question in providing adequate water for the use of lands requiring eight acres per cow month of

Location of water may be very important in range management.

grazing (somewhere near the present average of use on the public domain lands in the grazing districts managed by the Bureau of Land Management of the U.S. Department of the Interior); consequently the availability of natural water limits the season of use and the management of such lands. Some of these lands are now used even though the average travel distance to water is six miles—too far for anything but range horses and steers, and rigorous operation for the latter. Hauling water in tank trucks to supplement inadequate natural water on winter and spring-fall sheep ranges of the intermountain region is an increasing practice, and in some situations this has been remunerative in the increased production that has resulted.

Ranches of the plains and of the Rocky Mountain region have generally developed adequate water facilities, although ranches of the northern plains will no doubt develop more wells as a supplement to the impounding of surface waters, which may fail in dry seasons. A high proportion of the arid and semidesert lands of the intermountain and southwestern regions do not have adequate water for good range management. Water development is one of the most important things that can be done for the management of these lands and the enhancement of the production and income of the ranches. Such development will require close coordination between the ranch operations for private lands and the public land management agencies for adjacent and in many cases intermingled public lands, since this development must be wisely planned to be within feasible cost limits on the low-capacity lands.

In summarizing our observations concerning the distribution of grazing use by cattle over a range, we have observed that fences and the location of water development and of salt are complementary. All these need to be considered together in the plan. Where water supply is the limiting factor, fencing can sometimes be so planned that one water source serves two or more fenced range units. Then the location of salt at the opposite side (away from the water) of these units can aid in drawing the use away from the water. This applies especially to lands of reasonably favorable topography, where fencing can be planned to make good use of limited water sources. When rough terrain limits the use of fences on poorly watered ranges, the work of the rider becomes most important in achieving good distribution of use by cattle.

Securing good distribution of use on a range grazed by sheep depends to a considerable extent on what is practical in the use of a herder's wagon and well-spaced campsites, on the route of travel in covering the range, and on the ease with which the various units of a range can be covered from the campsite established for the unit. It is of especial importance in getting distribution of use over a sheep range to develop a plan of range units, based on the topography and the types and quantity of range feed, and to move the band from one unit to the next in a systematic plan of travel through the season of use. The role of the

Locating salt at strategic places between water helps to distribute the grazing use of the range. (U.S. Forest Service Photo)

Here is an illustration of open herding of the ewe-and-lamb bands on summer range. This is both good livestock management and good range management. (U.S. Forest Service Photo)

herder is very important in the success of a management plan for the use of a range by a band of sheep, not only in carrying out the plan put in making reasonable departures from it as the seasons vary.

PLANS FOR GRAZING LAND USE

Well-managed ranches have a systematic plan for the use of the grazing lands, even though such a plan may not be formalized and recorded. The information and analysis afforded by a range survey provide a valuable background for the development of such a plan.

Range Survey Information. As an aid in the preparation of the management plan for the rangelands of a ranch, a good range survey provides a map of the range resource and an accompanying set of descriptive notes. On the map are shown the main range forage types, the location of fences, water, and other improvements, and the topographic and elevation factors influencing the natural units or subdivisions for use and management of the range. As a supplement to the information provided on the map, the descriptive notes give, by sections and range forage types, observations concerning the composition, density, and vigor of the forage plants, and the changes that are occurring and probably have occurred in the range plants and soils. From the map and the notes a compilation is made of the acreages of the different major range types and of the estimated grazing capacity of each section or other acreages of range.

The range survey information, when carefully prepared and analyzed should show the possibilities and desirability of changes in the present management units and plan of use of the grazing lands. With this, it should show how development work such as fencing, water, reseeding, and the like would assist in the replanning of range units and making desirable changes in season of use, rotation management, or other practices. The range survey should give an ecological interpretation of any distinct trends that may prevail in range condition. The plant and soil information of the survey, when it is prepared in sufficient detail and recorded in permanent form, affords a basis for analyzing possible future changes in plant growth and range condition.

A Management Plan Based on a Range Survey. Figure 2, illus-

FIGURE 2. A MANAGEMENT PLAN FOR AN UPLAND SUMMER SHEEP RANGE

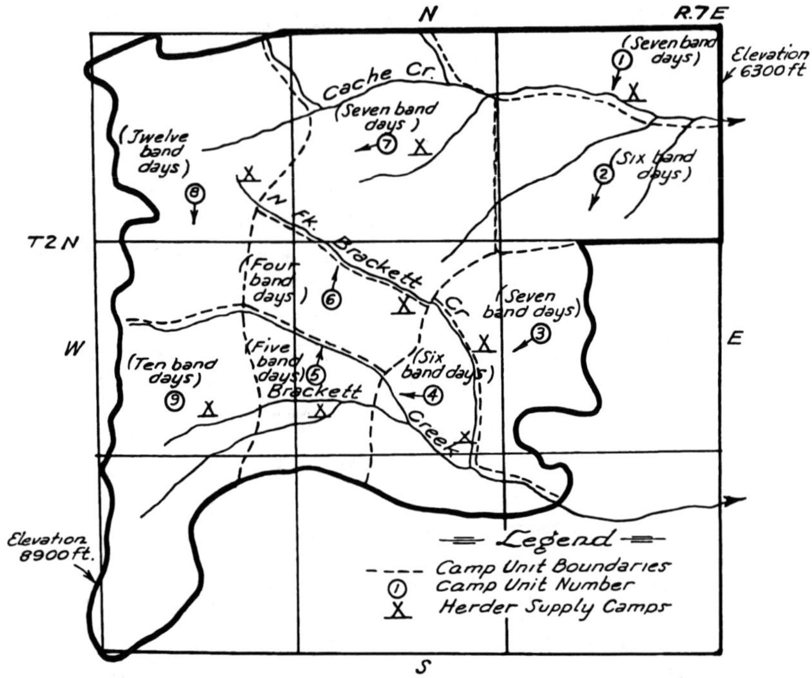

trating a management plan for the summer rangelands of a western Montana sheep ranch, was prepared on the basis of a grazing survey of these lands. Elevation of the land of the range area shown in the figure rises from 6300 to 8900 feet from east to west. Camp unit numbers show the sequence of use. Use is started on unit 1 between June 15 and July 1, depending on how advanced or late the season is. The route of travel for the band of 1500 ewes with lambs using this range is from north to south through the lower units, south to north through the intermediate units, and north to south through units 8 and 9, the high units.

This range furnishes about sixty band-days of grazing for the band of 1500 ewes and their lambs. This range is in the spruce-fir zone and is a mixture of grass and weed feed in the timber and the open parks, with about 40 percent of the feed within the timber stands.

Before the range survey was made and a management plan prepared, the method of use of this range was to move up the Cache Creek drainage during the first two weeks, and then to make heavy use of the extensive open parks in the high country of units 8 and 9. The return trail was down the Brackett Creek drainage, with about ten days of use in coming out. During the time this former method of using the range was in force, much of the early-season weed and browse growth of the lower country was lost, the feed in the timber of the intermediate and high country was lightly used, and the open parks of fescue grass and other green feed in the high elevations were heavily used to the point of depletion, with some invasion of unpalatable weeds.

With the present plan of management each of the camp units is used for a stated number of days by the band, the early weed feed of the lower country is used before it dries, adequate use is made of the feed in the timber of the intermediate country, and moderate use is then made of the high mountain parks (which now serve as a finishing range to produce as many fat lambs as possible). The initiation of this plan resulted in a substantial increase in lamb weights and in the percentage of slaughter-class lambs.

GRAZING LAND IMPROVEMENT AND DEVELOPMENT

Usually the purpose of rangeland improvement and development — reseeding, fencing, improving water facilities, contour furrowing, brush removal, rodent control, and the like — is to better the productivity and the management of a range. But the planning for such improvements and the valuation of their costs and benefits should be in terms of the management program of the entire ranch as a unit. As a consequence such plans should envision, first, any lack of balance that may now prevail in the various seasonal capacities of the range forage and feed crops and, second, the costs of and probable increased revenues from the improvements to increase the production of the ranch.

Some Cost Standards. By the price standards of the past, the upper limit of land investment for a cattle ranch of average productivity is about $7 per animal month of capacity. This is an average figure for all the deeded lands of the ranch, and it includes the value of the improvements for the use of the lands. But when a ranch with a moderate average land investment does not have sufficient range, for a short period of the year, to balance the other range and feed-crop capacities, it may be practical to invest $12 or $15 per animal month of capacity in the seeding of the range

and pasture lands needed to effect a better balance. As the practical cost limit in the seeding of rangelands and pastures in this situation, the total land investment of the ranch, including the cost of development work, should not rise above what can be justified by the productive capacity of the entire ranch as a well-balanced unit.

This same principle applies to the cost of other types of

Costly water developments must be planned to serve large areas. (Bureau of Land Management Photo)

range development. A program of water development to make better seasonal use of a range or to get better distribution of use by livestock on a range may, by effecting a better balance, increase the total productive capacity of the ranch considerably beyond the increased capacity made available by the development work. For example, a water development that makes available sixty additional cow months of feed on a spring-fall range may make it possible to carry an additional ten animals through the year.

Forecasting Costs and Benefits. Increasing the capacity of a ranch by range development and improvement can result either in increased market production from the same number of livestock or in an increase in the number of livestock that can be operated. Where the purpose of range improvement work is to compensate any major deficiency in the capacity of the rangelands of a ranch, the increase in productive capacity is more likely to come through an improvement in the weight and quality of market production per animal than through an increase in the number of animals operated.

Any plan of range improvement for the ranch should be based not only on an analysis of any prevailing lack of balance between the seasonal capacities of the ranch, but also on a forecast of the cost requirements for water development, reseeding, fencing, burning, and other types of work. The probable results of rangeland development and improvement, in terms of better market weights and market classes of animals, better calf and lamb crops, and other production factors, should then be estimated as closely as possible in terms of the probable increase in income, and compared with the estimated cost of the improvement work.

For example, a northern plains cattle ranch which operates 300 head of cattle has a deficiency of spring range and plans to acquire 500 acres of abandoned dry cropland to seed in crested wheatgrass. This land can be acquired for $1500 and the cost of seeding is estimated at $1200, making a total investment cost of $2700. It is estimated that the early spring grazing afforded by this acreage of crested wheatgrass will increase this ranch's annual market production of beef by 9000 pounds, an increase of 12 percent above the average annual production in the past. At an average price of $7 per hundredweight, the probable increase in annual income is approximately $630. On this basis, the investment cost of the land and the seeding would be recovered in about four and a half years. So even though we include an interest and tax charge on the investment in land and improvement and discount the future increase in annual income to a present value basis, the forecast indicates a substantial gain in net revenue.

MARGINAL RANGELANDS

When the average acreage requirement exceeds six acres per animal month for cattle grazing the cost of adequate range improvements for the management of the livestock and the land becomes increasingly prohibitive. Costs of adequate fencing and water developments for a township of grazing land (thirty-six sections) with an average grazing capacity of ten acres per animal month considerably exceed the capital value that can be sustained by the resource. As a consequence the improvements on such land are likely to be meager — below good management needs.

Land Use Margins. The necessary adaptation of cattle production to these management limitations on low-capacity desert and semidesert lands has been one of low production and low operating cost. Smaller and lower quality cattle are one means of adaptation to the additional travel for feed and water. Yearlong grazing and a minimum use of labor for livestock management are one means of keeping the operating cost consistent with the low market yield and income per head. The annual operating cost of this type of range cattle production is about half that on productive rangelands, and the gross income also averages about half that for the operations using good range. These low-production cattle operations cannot sustain much in land investment or land cost, and the rangelands they use, principally the remaining public domain lands, are used under permit or lease. Where these lands can be used in combination with higher capacity and better developed lands in ranch ownership, the value of the low-capacity land is considerably enhanced.

The economic limit for the use of low-capacity rangelands for range sheep production is below that for cattle, because it is practical to trail sheep considerable distances in order to use ranges during the season when feed and water are most available. The lack of fences is not a limitation on livestock management in the use of low-capacity lands by sheep, since they are under continuous supervision in the band. There are well-managed and productive range sheep operations using rangelands for which the acreage requirement on winter ranges is more than two acres per animal (ewe) month.

Private Ownership Margins. The economic margin for the use of low-capacity grazing land can be theoretically defined as the point at which the operation will sustain no land cost in leases, taxes, and interest return on land investment. In practice, before that point is reached there is a lowering of the return to the operator for his labor, management, and interest on equity in the investment.

The margin for the private ownership of low-capacity rangelands is above the margin for their use in livestock production, primarily because of the practical limitations of the present techniques for the appraisal of low-capacity lands and their valuation for taxation. Both the low-capacity and the high-capacity lands are valued too close to an average. Because of this, the cost of ownership of low-capacity grazing land is likely to be prohibitive even though its quality may be considerably above the economic rent margin.

It is largely for this reason that if any considerable part of the remaining public domain lands were to be moved into private ownership for grazing use they would have to be assessed for taxation at their productive values, $.50 to $.75 an acre. The poorer of these lands have little or no value. Transferring them to private ownership would not of itself enhance their productivity and value. The reason for the low production per unit of livestock on many of these lands is that their productivity will not justify the cost of good management facilities.

REGIONAL CONTRASTS IN RANGELAND MANAGEMENT

In the preceding pages we have seen some of the special regional applications of the principles of grazing land use. There are some other regional differences that merit description or reemphasis.

Fluctuating Grazing Capacities. Two contrasting viewpoints prevail regarding the use of a range capacity figure — necessarily an average for a period of years — for controlling the rate of stocking grazing lands. One is that the management program of the ranch should aim to carry a fairly constant number of livestock, commensurate with the average annual range forage production of a considerable period of years. The other viewpoint is that livestock numbers must be varied within rather wide limits over a period of years, increasing numbers to take advantage of favorable climatic trends, decreasing numbers to compensate for the lower forage production of unfavorable seasons.

These two viewpoints originated in different regions. They reflect important differences in the application of range management concepts. The idea that the livestock numbers of the ranch must be varied considerably over a period of time originated in the plains regions. The concept of holding the numbers fairly constant, based on the average grazing capacity of the rangelands for a series of years has been developed in the management of the foothill and upland ranges of the Rocky Mountain and intermountain regions.

Great Plains grassland ranges do undergo great fluctuations in their yield. There is a semblance of a trend in these fluctuations — a series of favorable years followed by a series of unfavorable years. But the trend may reverse abruptly, and the "average year" is pretty much a mythical quantity. Time and experience have demonstrated that the sod-forming grasses of the plains are very durable (except when plowed) in spite of the climatic change and the heavy grazing they may undergo. Where good management has been practiced since the severe drouth period of the thirties, the northern plains ranges have made a rapid recovery.

An economic balance between livestock numbers and forage production of the plains rangelands can best be achieved by adjusting to the climatic trends within such limits as will give good livestock market weights, and the effects of climatic change on range forage production can be modified somewhat by not using the full capacity of the range in the favorable years. The livestock production is penalized more than the resource for overusing the plains grasslands. Here the relationship between livestock production and the maintenance of the resource is fairly clear-cut and direct. But as a general rule, the range resources of the Rocky Mountain and the intermountain regions differ materially from the Great Plains in this very important feature of management.

Grazing capacities of the rangelands of the Rocky Mountain and intermountain regions are not generally subject to the wide climatic fluctuation and the rapid changes in plant growth that characterize the plains. There are some exceptions to this. One such exception is the great cheatgrass infestation of the sagebrush-grass ranges of the intermountain region. Another is the desert ranges of southern Arizona, southeastern California, and southern Nevada, which have a wide fluctuation in the growth of annual plants in the winter months. Grazing capacities of the ranges of the western mountain and foothill land do not vary widely from year to year primarily because of the variety of feeds — the perennial bunchgrasses, browse, and palatable weeds — that are available. To a somewhat lesser degree this generalization applies to the arid lowlands and the basins of the intermountain region.

For these lands a close relationship between a conservative "average year" grazing capacity and livestock numbers is not only practicable, but extremely important in a sustained management program for the ranch. This is of particular significance for these lands because it is possible for a period of several years to carry numbers considerably in excess of the "sustained yield" capacity of the range, and this can be done over a fairly long period *without any appreciable effect on livestock production and market weights,* while at the same time damaging and perhaps permanent changes are occurring in plants and soils. This is caused, at least in part, by the overuse of the perennial bunchgrasses and their replacement by shallow-rooted annual grasses or weeds that afford some feed, and by some shift to the use of the browse feeds by the livestock. This change has forced a drastic reorganization in the management program of a great many ranch operations, generally to the disadvantage of production and income.

Reseeding Depleted Ranges. Range science and ranch experience have in recent years provided considerable information on regional and local adaptations of range plants and cultural methods for reseeding depleted rangelands and abandoned plowed lands.*

Crested wheatgrass and (on good sites) smooth bromegrass seedings for early spring use are a valuable complement to the seasonal growth of the native grasses of the northern and central plains regions. Crested wheatgrass has so far proved to be the more valuable of these two grasses as a complement to the native grasses of the plains. It grows early, withstands heavy use during early growth, and still matures and makes seed during the late spring or early summer after use has been shifted to the native ranges. Crested wheatgrass is not a good range feed after it matures since it becomes woody and unpalatable, but there is often sufficient regrowth in the fall to afford some fall grazing.

Crested wheatgrass has shown considerable success on the better lands of the sagebrush-grass type of the intermountain region and the need of the northern parts of the intermountain region for spring and fall range (especially early spring range) is met in part through the seeding of crested wheatgrass. Also to some extent bulbous bluegrass seedings and rye pastures serve this same purpose. Those parts of the intermountain region that are deficient in upland summer range need adapted summer-growing grasses that are better suited to lowland pasture production than the native

* The technical literature on rangeland reseeding appears to have overemphasized increasing the grazing capacity of depleted rangelands rather than placing the emphasis on the use of adapted grasses to supplement and balance the seasonal use of native grasses. For a treatment of rangeland reseeding, applying particularly to the northern plains region and to the Rocky Mountain region, see *Reseeding to Increase the Yield of Montana Range Lands,* U.S. Department of Agriculture Farmers' Bulletin No. 1924, February 1943. For a similar treatment applying to the northern part of the intermountain region, see *Reseeding Range Lands in the Intermountain Region,* U.S. Department of Agriculture Bulletin No. 1823, July 1939.

Crested wheatgrass has shown good results for range reseeding in northern climates. (Soil Conservation Service Photo)

irrigated meadows that are now used to supplement late summer and fall grazing.

Timely Marketing of Livestock. Still another aspect of grazing land management with different regional applications is the timeliness of marketing cattle in relation to the trend in weight gains.* The *rate* of gain reaches a peak in early summer, as a rule, and starts to decline as the range feed begins to cure (the decline in rate of gain starts significantly sooner on a heavily stocked range). It has been customary in the northern and central plains to do most of the marketing in October and November, after the rate of gain has dropped to a low point or even when the market animals have begun to lose some weight. This low rate of gain in the fall may be offset some by a lower shrink during shipping than would occur in earlier marketing, but marketing a month earlier than has been the practice would often give a more efficient use of the range feed, both for the market animals and for the herd maintained on the ranch.

In contrast with this relationship between the time of market-

* For an analysis of this see *Market Your Range Cattle in Best Condition,* U.S. Department of Agriculture, AWI-55, July 1943.

ing and weight gain trends in the plains, it is customary for the cattle ranches of the Rocky Mountain region to move the market animals from the range onto the native hay meadows during September, thereby continuing the gain on green feed for another two to six weeks. This applies particularly to the ranches that are marketing some proportion of older animals — yearlings or two-year-olds.

However, the time of marketing as related to trends in weight gains and the efficient distribution in use of range feed between market animals and herd maintenance is likely to be a significant management factor for the cattle ranch throughout the West. This aspect of management appears to warrant special emphasis in the plains regions.

A LOOK AT THE FUTURE

The relationships between range plant growth, soils, climate, and grazing are complex. Sometimes the responses of range plants and of livestock production to these relationships appear to be contradictory and wholly unpredictable. It is, consequently, no surprise that a good many ranch operators have developed the fatalistic viewpoint that whatever changes may be occurring in range plants, soils, and livestock production are but natural uptrends

Rocky Mountain cattle ranches use meadow regrowth in the fall to get added gains and condition on market animals. (Bureau of Land Management Photo)

and downtrends. But some can see in these changes much more than a natural ebb and flow in range plant life. They see fundamental changes important to management.

These changes and their meaning in terms of management may not be easy to read. Frequently the help of a trained and experienced range technician is needed to comprehend these changes before the necessary management corrections have been too long delayed. This service of range management technicians to western stock ranches is, however, as yet largely in the future. Probably even the limited number of such technicians who now have the needed training and experience could not attain adequate recognition and employment as private consultants. In time this can and will change. The value of this service and the need for it *will increase*.

CHAPTER IV

Choice of Feed Crops for Ranch Use

GENERAL CONSIDERATIONS IN THE CHOICE

Most western stock ranches grow their own feed crops and produce only the amounts needed on the ranch. Usually the purpose in the use of crop feeds by the stock ranch is to maintain the breeding herd in thrifty condition and to meet the normal growth requirements of the young animals. Seldom does the experienced ranch operator try to make a large part of the weight gains by the use of crop feeds through the winter. Range livestock gains can be made far more quickly and economically on the range during the grazing season.

In the selection of feed crops best suited to the stock ranch the first question is what kinds can be grown economically. The second is which of these fit in best with the production program of the ranch. Some ranches are very restricted in this choice; others have a good many alternatives. Except in the locations where the spring runoff affords some early irrigation, native hay and dry-land feed crops constitute the alternatives for the plains ranches. Small-grain hay (wheat, rye, oats), sweet clover, corn, grain sorghums, and Sudan grass are the principal dry-land feed crops. Usually the plains ranch has some choice among these, though there are plains ranches with no alternative but wheat hay.

A majority of the ranches west of the plains can grow irrigated feed crops. Some of these ranches have ample irrigation water, but others have only early-season water in limited supply. Those with ample irrigation have the widest choice in selecting the feed crops best suited to the ranch. Alfalfa and other tame hays, native hays, and small grains afford the best possibilities for the ranches of the Rocky Mountain and intermountain regions. A considerable number of these ranches are still working out their adaptations of kinds and varieties of feed crops. An example is the work of some ranches with various mixtures of tame hays to get a better nutritional balance in the hay crop.

Some of the ranches of the Rocky Mountain and intermountain regions must grow their feed crops without irrigation or with only limited irrigation. Dry-land alfalfa, sweet clover, and grain hays are among the best possibilities for these ranches that lack irrigation.

SUITABILITIES OF DIFFERENT CROPS

Feed-crop production by the western stock ranch is not, as a rule, an intensive agriculture. Yields are not as high as those of the diversified farms in the western irrigation developments. Usually the objective of the ranch is to find the adapted type of feed crops that will meet the feed requirements with a minimum use of labor and equipment for production and harvest. In analyzing the opportunities of the ranch for feed-crop production it is, consequently, desirable to know the possibilities and limitations of the different types of feed crops that may be adapted to the climatic factors and the soil and water resources of the ranch. There are certain useful generalizations regarding the adaptability of the various feed crops to climates and resources, even though the individual

TABLE 3. PRODUCTION AND FEED CAPACITY PER ACRE FOR FEED CROPS USED BY WESTERN STOCK RANCHES

Crop and Kind of Land	Probable Yield per Acre (in tons)	Feed Rating	Feeding Rate per Month		Animal Months per Acre	
			Cattle	Sheep	Cattle	Sheep
Alfalfa hay						
Irrigated	2–3	excellent	500	120	8–12	33-50
Dry	½–1	excellent	500	120	2–4	9–18
Native hay						
Irrigated	¾–1¼	good	500	140	3–5	11–18
Dry	½–¾	good	500	140	2–3	7–11
Oat hay (dry)	¾–1½	good	500	140	3·6	11–21
Wheat hay (dry)	¾–1½	fair	700	175	2–4	8–16
Sweet clover hay (dry)	¾–1½	fair	700		2–4	
Corn (dry)	grazed off	good			4–8	
Sorghum (dry)	1–2	fair to good	600		3–7	
Sudan grass (dry)	1–2	fair to good	600		3–7	

ranch may have to do some experimenting to work out the most efficient production program for these crops.

Standards in Feed Yields and Capacities. Before appraising the regional and local adaptability of the various feed crops used by stock ranches, it appears desirable to consider some standards of production and feeding capacities of these crops, as shown by the operating records of a large number of ranches. The feed-crop yields in Table 3 are given as averages for the locations where the crop can be grown with reasonable success and regularity. Feeding rates used in this table constitute the full winter maintenance for a mature animal without other feeds or supplements. Feeding rates would be less than those shown in the table if the feeds were used along with grazing, or if the hay or roughage feed were supplemented with grain or some other concentrate feed. Feed requirements for wintering calves would be somewhat less than the average rates given for cattle. The higher feed requirements shown for the coarser and rougher feeds recognize the higher wastage in feeding, as well as the lower feeding values.

Alfalfa. This crop has the widest adaptability of any of the feed crops used by western stock ranches, from the Salt River valley of Arizona to the Milk River valley of Montana. This is due principally to the use of several varieties, which differ in winter hardiness, resistance to diseases, and ability to yield very well under good irrigation and reasonably well where only limited irrigation is possible.

It is accordingly desirable in the use of this crop to select the variety best suited to the climatic conditions and irrigation opportunities of the ranch. For example, the ranch that requires a winter-hardy variety and that has sufficient irrigation water to produce only one or two crops of hay a year would probably find Ladak to be a well-adapted variety of alfalfa. Only to a limited extent is alfalfa produced by western stock ranches without any irrigation, but there are many ranches that can do some irrigation with spring runoff or small local water storages. Where alfalfa can be grown under irrigation, even though water supplies are limited, the feed yield capacity averages higher than that of any of the other feed crops now used by western stock ranches.

Since alfalfa hay is relatively high in protein (12 to 18 percent), it is the best of the feed crops to use for limited hay feeding along

*Ranches that can grow alfalfa hay successfully find it their best field crop.
(Montana Engineer Photo)*

with winter grazing. This high protein content also makes alfalfa
a good winter feed for young animals that must have adequate
protein to meet body growth requirements. On the other hand the
mature animals, whose primary need is an energy feed for winter
maintenance rather than for body growth, can be wintered as well
or better on other crop feeds, such as native hay. The native grass
hay is, as a rule, considerably lower than alfalfa in protein, but
higher in the energy components — carbohydrates and fats. The
native hay that has an admixture of the better types of sedges is
likely to have the highest fat content.

Though alfalfa has the highest yield and feeding value of the
crop feeds produced by the western ranches, it also has a generally higher production cost per ton than the others. This fact needs
to be considered in making a choice among the feed-crop alternatives of the ranch. Few ranches have been able to achieve an average cost for alfalfa hay in the stack of less than $7 or $8 a ton,
when all costs are considered. One of the cost items often overlooked in the making of cost calculations is that of the periodic
renovation of the stands. Another is the annual cost of using the
land, that is, the land taxes and the interest charge on the investment in the land and water development.

Native Grass Hays. Native grasses are one of the important
hay crops of the ranches of the northern plains and the higher
mountain valleys of the Rocky Mountain and intermountain re-

Western wheatgrass on the bottom lands makes good hay for northern plains ranches. (Soil Conservation Service Photo)

gions. Yields of native hay are low compared with irrigated alfalfa yields, as may be seen from Table 3, but the cost of a native hay crop is correspondingly low.

Stands of native grass suitable for harvest as hay on the northern plains ranches occur mainly along the creek bottoms and the swales that receive some runoff waters. However, in favorable years there may be an opportunity to harvest hay on large acreages of dry rangeland. Because of the variability in the yield and production of native grass hay in the northern plains, it is desirable to accumulate considerable reserves in the stacks during the favorable seasons. The native grasses that are cut for hay — western wheatgrass, slender wheatgrass, and a variety and mixture of other grasses — will retain some feed value in the stack even though several years old.* There is, of course, a limit to the investment that the plains ranch should put into native hay reserves, even though it may be had in large quantities in favorable seasons for only the cost of harvest. This variability in the production of native hay can be met to some extent by the plains ranch through the use of low-cost methods of spreading runoff

* Some recent work on the vitamin content of range hay indicates that hay, especially hay in the stack, may soon become deficient in essential vitamins. Feeding some new hay with the old helps overcome the deficiency of the old hay.

Native meadows provide winter feed and fall grazing for high mountain valley ranches. (Bureau of Land Management Photo)

waters onto native grasslands, thereby securing greater dependability of yield.

As a rule the mountain valley ranches produce their native hay under irrigation, usually extensive and low-cost irrigation where the native meadowlands are flooded in the spring. This requires a minimum of land development for leveling and for the construction of ditches and laterals, and a minimum of labor for the management of the water. In general there is just the one flood irrigation and only one crop of hay is harvested.

This method of producing native hay under irrigation, where the meadows are flooded and sometimes kept flooded continuously for a period of a month or longer, often causes a gradual change in the composition of the native plants of the meadows. As a general rule the use of the irrigation water, in both time and amount used, should aim to maintain a good balance between the palatable native grasses and the sedges and rushes. Certain types of sedges, with their high carbohydrate and fat content, make a valuable complement to the native grasses in the hay crop. But when the sedges are greatly increased through excessive use of water the hay yields are lowered and the hay is not as well balanced nutritionally.

Some of the mountain valley ranches have had their meadowlands become predominantly a stand of unpalatable wire grasses and the less desirable sedges through continuous and heavy flood irrigation.

The predominance of native hay over alfalfa hay on the ranches in the higher mountain valleys has resulted mainly from the shorter growing season there and from the desire of the operators to avoid the higher seasonal labor requirement of a more intensive hay crop, such as alfalfa. This applies particularly to the larger ranches in the more isolated areas where seasonal labor for irrigating and haying is more difficult to obtain. Sometimes the lack of late-season water for producing a second or a third crop of alfalfa is the deciding factor in the production of native irrigated hay rather than alfalfa.

Small-Grain Hay. Small grains are an important hay crop for the stock ranches of some localities in the northern and central plains. Oats and wheat, produced under dry-land conditions, constitute the principal small-grain hay crops. Where cropland that will produce these crops with reasonable success is available, they have a feed yield per acre as high as any other dry-land hay crop or higher. This comparison is shown in Table 3.

Both oats and wheat must be cut in the soft dough stage of the grain if they are to serve their purpose well as a hay crop. They do not have keeping qualities in the stack equal to native hay or alfalfa. They do not, consequently, serve as well in accumulating feed reserves. Oats are generally rated considerably ahead of wheat as a hay crop for the plains stock ranch, but wheat can be grown more successfully in dry seasons, and there are localities in the plains in which the soils and climate make wheat the best and most dependable winter feed crop that can be had. This is especially true for the ranches of the northern plains that do not have creek bottom lands, swales, and other favored moisture sites generally needed to produce a sufficiently heavy stand of native grasses for hay.

Sweet Clover. This dry-land hay crop serves a useful purpose for the stock ranches that must produce their feed crops under adverse soil and moisture conditions. Sweet clover is a hardy, deep-rooted plant that will grow on poor soils and grow quickly when moisture is available. It is used in the northern and central plains and on dry mountain valley lands in the Rocky Mountain and intermountain regions. It has greater adaptability to alkali soils than most other feed crops.

The coarse and woody stems of sweet clover lower its feed value and cause a good deal of waste in feeding. Both the annual and the biennial varieties must be cut during their first bloom or they lose most of their value as a hay crop. The coarseness and roughness of sweet clover hay make it suitable mainly for cattle and horses. The resistance of the heavy stems to digestion may cause internal bleeding in the digestive tracts of cattle, especially when the growth is rank or the cutting is delayed.

Corn. Dry-land corn has become a widely used winter feed crop for cattle throughout the northern and central plains. It appears likely that the crop estimates of the U.S. Department of Agriculture have not yet shown the full significance of the use of this crop by the stock ranches of these regions. Where the ranches of these regions have cropland that will grow corn with reasonable success, it has a higher winter feed-producing capacity than any other dry-land crop.

The varieties of corn used as a dry-land feed crop are usually

Northern and central plains cattle ranches use dry-land corn as a feed crop.
(Soil Conservation Service Photo)

the open-pollinated varieties, since few hybrid varieties have as yet been developed that will mature under dry-land conditions within the growing season that prevails through most of the northern and central plains. The varieties that are used, such as Falconer and Northwest, will produce a good growth of stover in most years, and considerable ear growth in the favorable years. The harvested yield of the grain is usually between ten and twenty bushels per acre.

The usual practice in the use of corn as a winter feed crop for the cattle ranch is to turn the cattle into the fields in the fall after the corn growth is mature and dry. This gives additional growth and finish to the market animals for the ranch that markets them as long-yearlings or older animals. The ranch herd is then maintained on the roughage and the remaining ear growth through the winter months, as long as sufficient feed is available.

Some ranches have recently adopted the management practice of reserving the best fields or parts of fields, through temporary fencing or the use of electric-control fences, for the long-yearling animals that are to be marketed. These animals are turned into the corn just as soon as it is ready in the fall, to make as rapid gains as possible before marketing. Through this practice it has been possible, during the favorable years from 1940 through 1948, for some operators to market long-yearlings with an average weight of 900 pounds and to sell them as "good" grade slaughter animals.

Sorghums. This feed crop is grown in the plains regions, especially in the central plains, either as a row crop or in close stands. It is usually cut and stacked or used as an ensilage feed, but it is sometimes grazed. There is, however, considerable danger of losses from grazing when frost damage has caused a high prussic acid content in the plants. The second growth, which may develop after the forage has been harvested, is also usually very dangerous to livestock when grazed after being frosted. Sorghum requires about the same soil and climate conditions as corn for satisfactory growth. There is an increasing use of this crop for the storage of feed in pit silos.

Sudan Grass. This plant, an annual forage crop closely related to the sorghums, has considerable importance as a dry-land hay crop in the central plains. The stems are finer and tiller more frequently than those of the sorghums. Sudan grass is often grazed, but it has the same danger of poisoning as the sorghums when grazed after frost damage.

Crested Wheatgrass. This grass is becoming more important as a dry-land hay crop in the northern plains. When handled properly the hay is palatable and nutritious, but it must be cut within a limited period in order to realize its value as a hay crop. This stage of growth occurs after the spring growth is well advanced, but before bloom. If cut later than this, much of the value of the hay is lost. In contrast with crested wheatgrass the native western wheatgrass should be cut during the blossom stage, but it can be cut later and still make fairly good hay.

An advantage of crested wheatgrass compared with western wheatgrass and other native grasses as a dry-land hay crop is that in the usual season crested wheatgrass will produce sufficient volume for a hay crop on drier sites than the native grasses. Thus the ranch that is lacking in the favorable moisture sites required in most seasons for a heavy growth of western wheatgrasses may find that a seeding of crested wheatgrass on dry cropland will afford the best available hay crop.

Small-Grain Straw. Many of the small stock ranches and stock farms of the plains produce some small grain, usually wheat, as a cash crop. Where this is threshed in a stack, the straw is available as a feed. This also applies to the stock ranches of the Rocky Mountain and intermountain regions that produce irrigated wheat and oats as a nurse crop in renovating the alfalfa stands.

Wheat and oat straw have a low feed value (oat straw is rated superior to wheat straw), but beef cows and steers can be wintered successfully where they have access to stacked straw and the straw is supplemented with a limited amount of concentrate feed, such as one to two pounds of cottonseed cake per head daily. Where some alfalfa hay is available, it can be used as a limited feed to supplement the straw. Approximately five pounds of alfalfa hay is considered the equivalent of one pound of cottonseed cake as a supplement in using small-grain straw for wintering cattle.

Some ranches use straw for the winter maintenance of cows in order to conserve limited supplies of alfalfa or other hay feeds that are better suited to meeting the growth requirements of calves and yearlings through the winter months. It is possible to winter beef cows on straw alone during a mild winter, if the cows are in good condition at the start of winter feeding. This generally occasions considerable loss of weight, however, and the cows will likely require extra feed and care at calving time.

Farm Crop Feeds. Stock ranches that are near or adjacent to irrigated farming districts make considerable use of a number of feeds that are essentially by-products of the cash-crop production of the farms. Probably the best illustrations of this are the ranch operations bordering the large central valley of California. Here the stock ranches are organized to use the great reservoir of crop feeds made available by the irrigated farms from midsummer until the fall rains start the growth of range feed. The stock ranches of the Rocky Mountain and intermountain regions that use the farm feeds from adjacent irrigated farm districts use them mainly during the fall and winter months.

The more important of the farm crop feeds used by stock ranches are pea and bean straw, small-grain straw, sugar beet tops, and several types of crop aftermath pasturage. With the exception of the California ranch operations the farm crop feeds, where they are used by stock ranches, are generally a supplement to the crop feeds produced by the ranch. It is seldom that a stock ranch owns irrigated land in an irrigated farming district for feed-crop production, because of the value of this land for diversified farming or specialty crop production.

LABOR REQUIREMENTS FOR THE PRODUCTION OF FEED CROPS

As may be seen in Table 4, significant differences prevail in the labor requirement per ton of harvested production for the hay crops most commonly used by western stock ranches. Alfalfa and sorghum are the hay crops with the highest average labor requirement per ton in the stack. Native hay produced under irrigation has the lowest average labor requirement; grain hay and sweet clover are intermediate in their labor requirement.

The labor requirements for the production of feed crops, shown in Table 4, cover all labor for producing and harvesting the crop, but not for feeding the hay. The figures on labor requirements per ton of hay put into the stack are based on the average of the man-hours per acre given in Table 4 and the average of the yields per

TABLE 4. AVERAGE LABOR REQUIREMENTS FOR FEED-CROP PRODUCTION AND HARVEST

Crop and Kind of Land	Labor Requirements for Production and Harvest (in man-hours)	
	Per Acre	Per Ton of Average Yield
Alfalfa hay (irrigated)	18–20	8
Native hay		
Irrigated	5–6	5
Dry	3–5	6
Sweet clover (dry)	6–8	6
Sorghum (dry)	10–12	8
Grain hay (dry)	4–8	6
Corn (dry)	6-8	

acre given in Table 3. Labor-time requirements per ton of harvested hay run higher for native dry-land hay than for native hay produced under irrigation, even though there is little or no labor in growing the dry-land native hay crop. Several factors account for this. The principal reason is the low yield of the dry-land hay. Another is the fact that dry-land native hay is frequently cut in small acreages, along swales and creek bottoms. A third reason is that the usual practice in irrigating native hay meadows does not have a high labor requirement.

The labor requirement shown for corn is only for growing the crop, and no time requirement is given per ton of harvested yield, since the usual practice is to graze the corn in the field. If the corn were cut and shocked or stacked, the labor requirement would average about twelve hours per ton of stover harvested. This is the primary reason why most of the cattle ranches that grow corn as a feed crop use it directly for pasture in the field.

The labor requirement figures given in Table 4 were drawn from the operating records of stock ranches. They do not apply to western farm operations for the production of these crops, since farm operations, on both irrigated and dry land, grow these crops by more intensive methods, with higher labor requirements and higher yields per acre.

WATER REQUIREMENTS FOR IRRIGATED FEED-CROP PRODUCTION

Water requirements for the production of irrigated hay crops vary considerably, depending on how well the land has been prepared for irrigation, and on the amount of labor used for the management of the water during irrigation. Generally the tendency has been for the stock ranch to economize on labor rather than on water, with the result that water is not used efficiently by the standards of intensive crop agriculture.

A fairly good general average for the production of alfalfa by the stock ranch is three fourths of an acre-foot of water for each irrigation, but half an acre-foot is ample where the land and the irrigation system are well prepared for efficient use of the water. The ranch that has only an early water supply may flood the alfalfa stands with three fourths or more of an acre-foot of water as the only irrigation for the season. This will generally produce only one crop of alfalfa, although in some seasons there may be sufficient regrowth for a second cutting.

Stock ranches that produce two or three crops of alfalfa hay use, on an average, 1½ to 2½ acre-feet of water for three or four good irrigations. The efficiency in the use of the water that is applied varies greatly with different types of soil, however, as well as with the adequacy of the irrigation system and the management of the water. A tight and nonporous soil, or a loose, gravelly soil, will lower the efficiency of the water that is applied. Where the efficiency in the use of the water is low, as much as 1½ acre-feet of water is sometimes applied in a single irrigation. A 75 percent efficiency in the absorption and retention of the water that is applied is considered good.

In the usual practice of irrigating native meadowlands by one flood irrigation, the amount of water that is applied is, as a general rule, 1 to 2 acre-feet. But it is not unusual to apply 3 or 4 acre-feet, or even more, over a period of several weeks. If the water is available the meadow may be given a second flooding after one crop of hay has been cut in July or August. The second flood irrigation is usually for a short period, and the purpose is to put a regrowth on the meadows for fall pasturage.

FEEDING PERIODS AND FEED REQUIREMENTS

The objective of management in the production and use of feed crops is to meet the requirements of the average winter season and to carry sufficient reserves to be prepared for an adverse season. The management plan for the ranch with respect to crop

Rocky Mountain foothill and valley ranches plan for three to five months of continuous hay feeding. (U.S. Forest Service Photo)

feeds must, consequently, be based on past experience as to the averages and the extremes of the opening and closing dates and duration of the necessary feeding period.

Mountain valley ranches of the Rocky Mountain region have the highest crop feed requirements, averaging about 1¼ tons of hay per head for cattle and 500 pounds per ewe. The length of the feed period averages about 5 months for cattle and 4½ months for sheep. The feeding period generally falls between December 1 and May 1. There are some very significant local differences in the feed requirements of the mountain valley ranches, however, because of local variations in winds, snowfall, topography, and the like. The foothill ranches of the Rocky Mountain region have an average winter feeding period of about 4 months, usually between December 15 and April 15. The hay feed requirement averages approximately a ton to the head for cattle and 450 pounds per ewe.

Generally the feeding period for the ranches of the northern plains falls between January 1 and April 1, with the probability that the livestock will be on a full maintenance feed of crop feeds for thirty to sixty days during this period, and on limited feed with some grazing during another thirty to sixty days. The average feed requirement is about half a ton of hay per head of cattle and 200 pounds of hay per ewe. These same standards also apply rather generally to the ranch operations of the Nebraska sandhills and the central plains. Because of their limitations in grazing lands, the stock farms of the central plains maintain the livestock continuously on crop feeds for a period of three to five months between November 1 and April 1.

Sheep ranches of the intermountain region do very little winter feeding, except for the Idaho and Oregon early-lambing operations of the Snake River valley. As a rule these early-lambing operations practice continuous and heavy feeding from about December 1 through March. The cattle ranches of the intermountain region that are in the lower semidesert country have to do only occasional and limited winter feeding. Those located in some of the higher intermountain basins and near the higher mountain ranges have a winter feeding period of two to three months, usually within the period from December 1 to March 1.

THE PLACE OF FEED CROPS IN THE MANAGEMENT PLAN
OF THE RANCH

In planning the production program of the ranch the various possibilities for feed-crop production and use should be analyzed as supplementary to, rather than coordinate with, the management requirements of the grazing lands. This point of view, that feed-crop production and use should be primarily to facilitate the management program of the grazing lands, is desirable because efficiency in the use of the grazing lands has a far greater influence upon livestock gains and production than feed crops. Furthermore, grazing land management requirements are more definitely fixed by natural factors; the croplands present more alternatives for control and adjustment than do the rangelands.

It may be said, consequently, that the management program for the use of the grazing lands should be the controlling factor in the choice among the various feed-crop production opportunities of the ranch, even though the cropland capacity may be a large proportion of the total capacity of the ranch. An illustration of this is the management situation of many mountain valley ranches in the Rocky Mountain region that have the cropland and water resources to develop an irrigated cropland capacity that overbalances the capacity of the grazing lands. Frequently the economic balance in land capacities for these ranches is realized by seeding part of the cropland to the locally adapted irrigated pasture grass or mixture of grasses for use during the season or seasons of the year that will best supplement the use of the native grazing lands.

Another illustration of an adjustment in the use of the croplands to facilitate the management of the rangelands is the grazing of irrigated native hay meadows in the early spring by some cattle ranches of the intermountain region. This may cause some sacrifice in hay yields, but the consequent shift from an unseasonably early use of the spring range can increase rangeland capacity much more than the possible loss in hay production.

Sometimes in the plan of organization and development of a ranch property the management planning has centered on the croplands, with an attempt later to fit grazing land capacities and seasons to the plan of use for the croplands. This will work on the stock farm, where the use of grazing land is minor, but seldom gives a satisfactory result for the stock ranch.

CHAPTER V

Markets, Prices, and Incomes

BASIC FACTORS IN WESTERN CATTLE MARKETING

Western Beef Production and Consumption. On an outline map of the United States, draw a line from north to south through Great Falls, Montana; Grand Junction, Colorado; and Las Cruces, New Mexico. Owing to the growth of West Coast cities during the past twenty years, beef consumption now equals or somewhat exceeds production in that part of the United States west of this line.

Table 5 presents this picture in a somewhat different way. These 1945 data show the seven far western states to be a deficit area, which must reach eastward for slaughter cattle and for beef. The four Rocky Mountain states produce a much greater total tonnage than they consume, but about half of this production is accounted for by the feeder cattle that move eastward to farm feedlots in the Corn Belt.

Effect of the Deficit upon Western Markets. The north and south line on your map marks fairly well the transition zone between the eastward and westward market movement of cattle. This is, of course, a general picture; some noteworthy exceptions prevail. A sizable number of young feeder cattle move eastward to the Corn Belt from west of this line. Fed cattle move from the irrigation-farming feedlot-fattening centers of the Rocky Mountain states — the Billings and Greeley districts, for example — to West Coast slaughter markets. Some California cattle ranches buy their stocker cattle as far east as western Texas. The burgeoning California cities are now reaching as far east as the processing plants of Omaha and Kansas City for the better grades of beef. Seldom, however, do cattle suitable for slaughter move eastward across the line drawn on your map.

TABLE 5. BEEF PRODUCTION AND CONSUMPTION IN THE
ELEVEN WESTERN STATES, 1945

State	Beef Production (dressed weight, in thousands of pounds)	Beef Consumption (dressed weight, in thousands of pounds)	Surplus or Deficit of Beef Supply (in thousands of pounds)
Arizona	90,000	44,100	45,900
California	325,000	617,610	−292,610
Idaho	120,000	35,000	85,000
Oregon	145,000	84,420	60,580
Nevada	48,000	11,200	36,800
Utah	65,000	43,190	21,810
Washington	120,000	146,930	−26,930
Total, far western states	913,000	982,450	−69,450
Colorado	235,000	78,470	156,530
Montana	225,000	32,150	192,850
New Mexico	145,000	37,450	107,550
Wyoming	130,000	17,290	112,710
Total, Rocky Mountain states	735,000	165,360	569,640

This picture tells a story of the difference between the western
cattle markets and those to the east of the plains. The indications
are that the cattle feedlots of the eleven western states as a group
do not turn off enough fed cattle to meet the demand of these
states for good quality beef. (Total dressed-weight tonnage of
feedlot-fed cattle normally amounts to about 25 percent of the ag-
gregate consumption of these states.) This limited supply of feedlot-finished cattle in the central markets of the eleven western
states (a general picture, there are exceptions) causes some contrasts with the midwestern markets in cattle classes, grades, and
prices. There is a strong tendency for lower quality cattle to sell
in a slaughter class and for slaughter animals to sell a grade higher
at the West Coast central markets than the same animals would
sell at Kansas City or Omaha. The western markets are normally more receptive to short-fed old cows and steers, hay-fed or
"warmed up" animals, range-fat cows and steers, than are the
markets at Kansas City, Omaha, and Chicago.

Some Probable Trends. Look again at the figures given in Table
5. The beef production figures are high, the result of a large cattle
population and a series of good seasons. Probably this beef production represents a maximum or even more than can be maintained unless there is a considerable further expansion in western

irrigation. On the other hand, the human population trend is sharply upward for the Pacific states. The beef consumption figure given in Table 5 for 1945 represents wartime rationing. The 1945 per capita average (for beef and veal) of about seventy pounds was high but may in the future run even higher. What does all this mean to the western range beef cattle grower?

Western market demand for good slaughter cattle is likely to increase. Feedlot finishing in the West depends mainly on irrigation farming, and though the present irrigation developments can probably expand their feeding, any large increase in western feeding must wait on an expansion in irrigation.

Meanwhile, the western markets are going to bid for animals that come from ranges and pastures in good condition. The western grower that can manage his ranges, pastures, meadow grazing, and feed crops to produce a good beef animal is likely to find a better western market in the future than in the past (the war period excepted). Management of the summer ranges and fall pastures to serve as "finishing ranges" for the market animals may take on added significance for those western range cattle producers that have such possibilities.

Increased demands of western markets for slaughter cattle may be in some degree offset by the purchase of dressed beef from midwestern processing plants. Freight rate relationships between the westward movement of dressed meats from midwestern processing plants and westward shipments of slaughter cattle from the Rocky Mountain states will have considerable bearing on the future trend of western markets for the cattle ranches of these states. Freight rates favoring the westward movement of dressed beef over slaughter cattle could be a strong influence causing western states to market more young feeder cattle eastward, fewer mature slaughter animals from ranges and pastures to the western central markets.

MARKET CLASSES AND GRADES OF CATTLE

The Basis of Classes and Grades. The market classifications* of cattle that signify the market use of animals are (1) slaughter cattle, (2) slaughter veal calves, (3) slaughter calves, and (4)

---

* For a complete treatment of market class and grade standards for cattle and sheep, see *Market Classes and Grades of Livestock,* U.S. Department of Agriculture Department Bulletin No. 1360.

feeder and stocker cattle and calves. Most of the western range marketings are in the first and fourth of these classifications. These market classifications are further subdivided by sex, age, and weight of animals, and it is to these subclasses that the market grades apply. For example, in the market classification for slaughter cattle, the grades applying to the various age and weight classes of steers are prime, choice, good, medium, common, cutter, and canner.

The general market class of slaughter cattle covers all animals intended for immediate slaughter and more useful for that than for any other purpose. The general market class of feeder and stocker cattle and calves applies to the animals that show evidence of ability to grow and take on additional weight and finish, and that are used or will be used for feeding and fattening purposes. An animal that goes as a slaughter animal might, under a different market supply and demand situation, go as a feeder. Cattle that may go to either the packer buyer or the feeder buyer are often referred to in common market parlance as "two-way" cattle.

The terms *feeder* and *stocker* are often used synonymously and in combination. However, stocker cattle are generally those that are used for further growth and weight development on pasture and roughage feeds. Feeder cattle are those that are used for rapid fattening by intensive feeding. Market grades applying to the feeder and stocker steers and to heifers, calves, yearlings, and two-year-olds are fancy, choice, good, medium, common, and inferior. The grades applying to cows in the feeder and stocker class are choice, good, medium, common, and inferior.

A rather large proportion (it varies by seasons and years) of the range cattle marketed go as slaughter animals. This is necessarily the situation because the total number of cattle in the feedlots through the year usually is, nationally, a fourth or less of the total number slaughtered (excluding calf slaughter). Because of this considerable sale of range and pasture animals for slaughter and the sometimes large price spreads between the "common," "medium," and "good" grades, it is important to the western grower that his slaughter animals rate as well up the scale in grade as possible.

Factors That Determine Grades. The factors that determine the grade of an animal in its market class are conformation, quality, and finish. An animal with a desirable conformation is well proportioned in width, depth, and length of body, and it has a good distribution of fleshing. Conformation is determined principally by the breeding and finish of the animal. The quality of an animal depends primarily on breeding, age, and finish. Finish refers to the fleshing and fatness of the animal and is important in relation to the dressing percentage or yield and the quality of the meat. It is unusual for unfed western range steers or cows to grade higher than "good" in the slaughter class. Feedlot finishing is generally necessary to produce the "prime" and "choice" grades.

A high proportion of the western cattle that sell as feeders are calves and yearlings. Ordinarily the feedlot feeder does not want the older cattle that have already made most of their growth. The mature cattle, if in good flesh, are more likely to go as slaughter animals; they may sell as stockers if they are thin but apparently capable of making good gains on pasture and roughage feeds. Thin cattle that are mature or nearly so are likely to sell in one of the low slaughter grades in the absence of a demand for stockers.

CATTLE MARKETING METHODS

Consignment Shipping and Cooperative Marketing. The marketing practice most generally used in the past by the western range cattle grower has been to ship to one of the central markets on consignment to a commission merchant for his sale. Then the commission merchant is the agent of the grower in finding the best market outlet to the packers, feeder and stocker buyers, or other types of buyers.

During the past two decades cooperative organizations of growers have become more important in the marketing of western cattle. The cooperative organization may be a marketing cooperative with facilities and connections for selling through the central markets or other market outlets (such as directly to feeders), or it may simply be a shipping cooperative that markets through the central markets, in the same manner as an individual grower, and handles the accounts for the growers. The principal service of the shipping cooperative is that of making up carload lots for shipment by the smaller growers. It is not very important in western ranch marketings.

Direct Sale to Packer and Feeder Buyers. Direct sale by the grower to the feeder buyer is an important market outlet, particularly in the plains regions. The feeder buyer often buys at the ranch, but after satisfactory connections have been established the transaction may be handled in succeeding years as a "mail-order" deal between the same grower and feeder.

Direct sale by the growers to packer buyers is important in some sections of the West, particularly in the Pacific Coast states. Probably the West Coast packers have expanded their direct buying contacts in recent years. Direct sales to packer buyers usually occur at country points, but in some cases contact between the grower and a packer buyer takes place at a central market.

Country Dealers and Auction Markets. The country dealers who buy on their own speculative account or "on order" for feeder and packer buyers are not as important a factor in market organization as they formerly were. One reason for this is the recent development of the country auction market. This has become an important outlet for western growers, particularly for the small growers. The auction market is usually organized and operated as a local enterprise. Auction markets provide yard and auction service on a fee basis and help to give the grower something that has often been lacking in country sales — adequate competition between the country buyers. They usually know current market values better than the grower.

MARKETS FOR WESTERN RANGE CATTLE AND MARKET MOVEMENTS

Approximately two thirds of the cattle produced by the western growers move through the principal central markets. This includes shipments of growers, cooperatives, and country dealers. The other one third is sold to local outlets — to the smaller local markets, direct to the packers, or direct to feeder buyers outside the range territory. There are some limited market channels that do not go through the central markets from the points of western origin. Country buyers who buy stockers and feeders "on order" for clients outside the ranching country are an illustration of this.

Principal Central Markets. Kansas City and Omaha are the central markets outside of the range territory that receive the largest numbers of western range cattle. The others of importance are Sioux City, St. Paul, Chicago, Wichita, and St. Joseph. The most important central markets of the western states are Fort Worth, Denver, Oklahoma City, Los Angeles, Portland, Ogden, San Francisco, Salt Lake City, Spokane, and Seattle. The movement of range cattle to central markets from North and South Dakota is principally to the St. Paul, Sioux City, Chicago, and Omaha markets. Omaha, St. Joseph, and Chicago are the important central markets for Nebraska cattle, Kansas City, St. Joseph, and Wichita for Kansas. Cattle from the Texas ranches move mainly to Fort Worth and Kansas City, with about 15 percent going to California markets. St. Paul, Chicago, and Spokane receive most of the Montana marketings. Denver and Omaha are the principal central markets for cattle from Wyoming and Colorado. About half the New Mexico movement to central markets goes to California, half to Denver, Fort Worth, and Kansas City.

There is little or no movement of cattle from California to central markets outside the state. The California markets take practically all the export from Arizona and Nevada, and about half the movement to central markets from Utah. The other half of the Utah central market movement is to Ogden and Salt Lake City. The Idaho movement to central markets is principally to the Utah markets and to Portland and Seattle. One third of the central market movement from Oregon is to California markets and two thirds to Portland. Spokane, Seattle, and Portland are the principal central markets for Washington.

Marketing Seasons. Range cattle from the northern plains and Rocky Mountain regions move to market mainly during the months of September, October, and November. There is no such distinct autumn peak in the market movement from the central plains region, because of the stock farm character of the range production and the large movement of stocker animals into this region during the fall and winter months. The market movement from Texas runs about even through the summer and fall months. The cattle of the intermountain region move to market principally from August through December. California marketings reach their peak in May, June, and July.

WESTERN LAMB PRODUCTION, CONSUMPTION, AND MARKETS

Western Production and Consumption. Five of the seven far western states produce considerably more lamb than they consume. In 1945 the California lamb consumption exceeded production by some 11 percent, that of Washington by some 30 percent. However, Table 6 shows that these seven states as a group produce a large surplus of lamb.

TABLE 6. LAMB PRODUCTION AND CONSUMPTION IN THE
SEVEN FAR WESTERN STATES, 1945

State	Lamb Production (dressed weight, in thousands of pounds)	Lamb Consumption (dressed weight, in thousands of pounds)	Surplus or Deficit of Lamb Supply (in thousands of pounds)
Arizona	9,500	4,410	5,090
California	55,220	61,660	—6,440
Idaho	43,000	3,500	39,500
Oregon	20,600	8,442	12,158
Nevada	9,165	1,120	8,045
Utah	30,500	4,319	26,181
Washington	11,250	14,613	—3,363
Total	179,235	98,064	81,171

The production and consumption relationship shown in Table 6 for these seven states gives a valid general picture. It should be noted, however, that the production figure is somewhat unrealistic because it includes a dressed-weight calculation for the feeder lambs shipped eastward from these states. However, the dressed-weight yield of the range-fat lambs and the feedlot-fed lambs of these seven states for 1945 is estimated at 123 million pounds. This is still considerably above the 1945 consumption shown in the table.

Western Markets. We see from the analysis in the table that the market flow of western lambs, both feeder and slaughter lambs, is eastward from the states west of the continental divide. The minor California and Washington deficits do not materially modify this eastward flow. Because of the highly seasonal California spring-lamb marketing, much of the California production does move eastward; but this is more than offset by the import of fed lambs and fall-season range-fat lambs. The Rocky Mountain region and the plains regions all produce a large surplus of lamb. In contrast with the market flow of western cattle, there is no transition zone in the eastward and westward market movement of lambs.

The general market movement of western range lambs is, then, to the midwestern feeding centers and to the eastern consuming centers. From 60 to 70 percent of all western range lambs sell as feeders (all those in the plains), and in the usual season about one third of these go to lamb feeders in the western irrigated districts. Since a much larger percentage of western range lambs must sell as feeders than is true for range cattle marketings, the supply and the price of feeds are somewhat more important in determining lamb prices.

Western per capita consumption of lamb has in the past averaged rather low, especially in comparison with the important eastern consuming centers (mainly in the Middle Atlantic and the New England states). A continued rise in the West Coast population centers and a moderate increase in their per capita consumption of lamb could conceivably reverse the present eastward movement of slaughter lambs from the states west of the continental divide.

MARKET CLASSES AND GRADES OF LAMBS

The Basis of Grades and Classes. The market classifications of lambs as to use are (1) slaughter lambs and (2) feeder lambs. A young sheep is classed as a lamb until it is approximately one year of age, when it acquires its first pair of permanent teeth. Between this time and the appearance of the second pair of permanent teeth at about two years of age, the animal is classed as a yearling sheep.

General age groups commonly used in market practice for slaughter lambs are (1) hothouse lambs, (2) spring lambs, and (3) lambs. The hothouse lamb usually is less than three months of age and weighs 30 to 60 pounds when marketed. The growth of the hothouse lamb is made by feeding the ewe for heavy milk production. This type of slaughter lamb is a farm rather than a range product.

Spring lambs are those that come from ranges and pastures in slaughter condition. Spring lambs go to market during April and May in California and Arizona, and through the summer and fall months in the other western states. Spring lambs are generally five to seven months of age when marketed. The term *lamb*, in the slaughter class, applies to feedlot-finished lambs; these are generally nine to twelve months old. Market grades applied to slaughter lambs are prime, choice, good, medium, common, and cull. Usually the market weight of spring lambs, as slaughter lambs from the range, is 80 to 90 pounds. That of fed lambs generally runs between 85 and 95 pounds.

Feeder lambs, which make up the bulk of western range lamb marketings, usually weigh between 60 and 70 pounds when marketed. The market grades applying to feeder lambs are fancy, choice, good, medium, common, and inferior. Most range feeder lambs are six to eight months old when marketed.

Factors Affecting Classes and Grades. The lamb production of the western grower who uses mountain summer range is often a combination of feeder- and slaughter-class lambs, because of some difference in the age of the lambs and the natural differences in individual thriftiness and growth rates. It is of distinct value to such a grower in dealing with country buyers if he can judge at marketing time the approximate division of his product between the slaughter and the feeder classes. Not many growers are able to do this.

Western range growers are usually able to produce the better grades of feeder lambs in large lots, thrifty and of uniform age and weight, averaging about 65 pounds. The range growers who reach the top grades in slaughter lambs are, as a rule, those who use the mutton-breed sires with ewes of the fine-wool breeds, or mutton-breed sires with crossbred ewes, and who have an exceptionally good upland summer range.

MARKETS FOR WESTERN RANGE SHEEP AND LAMBS AND MARKET MOVEMENTS

Marketings of lambs by the western growers are about equally divided between (1) selling through the central markets by the individual grower and by cooperative organizations of growers, and (2) the country sales to feeder buyers, packer buyers, order buyers, speculator buyers, and through the country auction markets. Lambs make up the bulk of the marketings. Normally about nine tenths of the growers' sheep sales are lambs. Marketing of ewes is largely through the central markets, though in some years there is considerable direct sale of ewes to western farm-flock operators by the range operators.

The Central Markets. Omaha, Kansas City, Chicago, St. Joseph, St. Paul, Sioux City, and Wichita are the most important of the midwestern central markets for range sheep and lambs. The important central markets of the far western states are Denver, Ogden, Fort Worth, and Salt Lake City. Most of the market movement from the northern plains to central markets is to St. Paul, Chicago, and Sioux City. Principal central markets for the central plains are Omaha, Kansas City, and Wichita. Texas marketings are largely to Fort Worth and Kansas City. A high proportion of the movement from the northern part of the Rocky Mountain region goes to Chicago and St. Paul. Denver and the Utah markets are the principal central markets for the central and southern parts of the Rocky Mountain region. For the intermountain region the central markets are Denver and the Utah and California markets, with some movement to the midwestern markets, principally Kansas City and Omaha.

California, which is a net importer of lamb, markets two thirds of its production to the Denver, Utah, and other western markets outside the state, and imports from Arizona, Utah, Nevada, and Oregon. This is caused by the highly seasonal marketing of the California spring-lamb crop — 70 percent of it in May — and the comparatively small number of lambs fattened in the feedlots of California central valleys, approximately 125,000 head out of the annual California market production of 1,200,000 to 1,500,000.

Seasonal Movements. With the exception of California, Arizona, and Texas, the seasonal movement of lambs from the western growers is principally in the fall months, largely during September, October, and November. California and Arizona marketing reaches a peak in April and May. There is an early season movement of grass-fat yearling sheep from Texas in April, May, and June. Range lambs begin to move from Idaho, Oregon, and Washington in July, earlier than the start of marketing from the Rocky Mountain and intermountain regions.

WOOL MARKET GRADES

The Basis for Grades. Wool grades are based primarily on the fineness and length of the fibers in the fleece. The grades most commonly used at present in the grading of wool in the United States are fine, ½ blood, ⅜ blood, ¼ blood, low ¼ blood, and braid and common.

These grades range from a fineness in diameter of fiber, short length of fiber, and high "crimp" in the fine wools, to coarser and longer fibers and lower crimp in the coarse wools. As a rule the finer grades come from the sheep breeds that have a high proportion of Merino blood. The wool grade designated as "½ blood" is, in theory, wool with the characteristics of fineness, length, and crimp of fiber that would come from a sheep with ½ Merino blood.

Most of the wool from western range sheep is finer than ⅜ blood. The bulk of the western range production grades ½ blood and fine. Most of the wool from the western states grading ⅜ or coarser comes from the ranches that emphasize slaughter-lamb production and, consequently, use ewes that are a cross between the fine-wool breeds and the mutton breeds. Wool from the western farm-flock sheep of the irrigation farming districts usually grades ⅜ or ¼ blood. These operations, like those of the eastern or "native" sheep states, use Suffolk, Hampshires, or other mutton breeds, or crossbred ewes. Braid and common grade wool is unimportant in western production. This grade is produced by the Lincoln and Cotswold breeds.

Grades of wool may also be expressed in terms of the "spinning count" of wools with different fineness and length of fibers. A wool that grades fine staple (other grades of fine wool having a shorter length of fiber than the fine staple grade, which must have a minimum fiber length of three inches, are recognized) has a spinning count in the 64's, 70's, and 80's. That is, with a spinning count of 64, one pound of scoured and combed wool can be spun into 64 hanks of worsted yarn (560 yards of yarn length to the hank). Half-blood wool has a spinning count in the 58's and 60's; ⅜ blood in the 56's; ¼ in the 48's and 50's; and low ¼ blood in the 46's. The spinning count for braid and common wool is in the 44's, 40's, and 36's.

Significance of Grades. These wool grades in terms of relative fineness and length of fiber do not signify differences in the inherent quality of wools. Rather they designate the suitability of wool for different types of use. The fine grades of wool are used for women's fine apparel and the finer weaves of cloth for men's suitings. The grades from $\frac{1}{2}$ blood to $\frac{1}{4}$ blood are used in the manufacture of worsted cloth for men's suitings. The coarsest grades of wools go principally into carpet manufacture. There is little of this wool produced in the United States.

The relative demand and price for the supply of the different grades of wool available in the United States vary considerably from year to year with changing styles in clothing, the extent of the use of synthetic fibers for certain types of clothing manufacture, and the general level of incomes and purchasing power of consumers. It is not, consequently, possible to see any distinct trend in market demand either favorable or unfavorable to the finer grades of wool that the western range growers produce. Rather than changing the breeding program of the ranch, shifting wool production toward a coarser or a finer wool in response to wool price changes, the western ranch should aim to produce the best grade and quality of wool for the breed and type of sheep adapted to the ranch.

The quality of wool within a grade depends upon the uniformity in diameter of the fibers, the strength of the fibers, and the cleanness of the fleece. Uniformity of the fibers, both in the individual fleece and in the clip, is attained mainly by following a consistent program in the culling and selection of ewes and of the breed and type of sires. The strength of fibers as a factor affecting wool quality depends principally on an adequate level of nutrition through the winter months. Cleanness of the wool as a quality factor depends on the absence of burs, seeds, manure, and other foreign substances that may stain the wool or that are difficult to scour out of the wool.

WOOL MARKETING

Marketing Methods. The present methods for marketing western wools are the following: (1) consignment to wool commission merchants who warehouse the wool and sell it to the mills, (2) grower cooperatives that have terminal warehouse and sales facilities, (3) sales to wool dealers who buy on their own speculative account or on order for mills, and (4) direct sales by the growers to the mills. The advantage to the grower of the first two methods is that the wool is in the hands of a marketing specialist who knows grades and values and who can show wool to prospective buyers at a central market point, thus attempting to get the highest competitive price for the grower.

A Limitation of Present Methods. A disadvantage of much of the present wool marketing is that there is not an adequate measure of the differences in values of wools of different shrinkage and other quality factors, and the grower consequently does not have the information on his clip that he needs to guide him in management. The reason for this is that many wool sales are made on the basis of a judgment appraisal of the grease wool, rather than on the scoured sample, to determine actual clean-weight yield and quality of the clip. Appraisals, however good, tend to cluster around an average. An illustration of this for the shrinkage factor of wool quality is given by the following tabulated comparison of two western clips. These clips were appraised as having about the same shrink and other quality factors and were sold at nearly the same grease-weight price per pound.

Clip Number	Average Grease Weight of Fleece (in Pounds)	Price per Pound to Grower	Average Return per Fleece	Percent of Shrinkage	Clean Wool Yield per Fleece (in Pounds)
1	9.3	$.454	$4.22	56.4	4.05
2	8.4	.467	3.92	52.4	4.00

Both these clips graded ½ blood and were the same type of wool. This illustration of one clip selling for an average of thirty cents per fleece more because of the additional dirt it contained is not at all extreme. The limitation of wool appraisal may be overcome fairly soon by the development of a sampling technique to determine the clean-weight yield of grease wool. This technique is "core sampling" of the bags of wool, using a mechanical device to cut a core of wool out of the bag of wool. This core is than tested for shrink and grade factors.

The dirt content of wool is not, of course, entirely controllable through management. Winter feeding on snow, for example, will produce a cleaner fleece than winter ranging on desert land. But the cleanness of the wool may be influenced by management methods such as the selection and frequency of change of bedgrounds. When the service is available to growers for the marketing of wool on an actual sample test of the shrink and the quality factors of the wool, they will have a better guide and incentive for management in the improvement of wool than is afforded by present wool appraisal methods.

SHEEP AND WOOL PRICES AND RANCH INCOMES

Prices. Prices received by growers for sheep, lambs, and wool during the twenty-year period from 1921 to 1940 have varied 50 percent or more above and below the average for the period. Let us see what medium high, medium, and low prices mean in terms of probable gross income for different types of western sheep ranches. In selecting the high and low price points we will not use the extremes that have occurred; rather let us take those that appear more usual and probable in the price variations from high to low. The prices used in Table 7 are the averages of prices received by western growers, as shown in U.S. Department of Agriculture price statistics. There are important regional variations in these prices due to the distance from market; there are individual ranch variations due to quality of product.

TABLE 7. MEDIUM HIGH, MEDIUM, AND LOW PRICES TO GROWERS FOR
WESTERN LAMBS, WOOL, AND SHEEP

	Prices per Pound to Growers		
	Medium High	Medium	Low
Lambs			
"Good" grade feeders	$.11½	$.08	$.04½
"Good" grade slaughter lambs	.13	.09½	.06
Wool	.40	.27	.15
Aged and cull ewes	.06	.04	.02

SOURCE: These figures are based on the price statistics of the U.S. Department of Agriculture.

Using Prices to Estimate Probable Gross Income. The prices shown in Table 7 can be applied to the market production of the ranch to estimate the probable gross income of a ranch under different price conditions, and these calculations can be made on a per-head basis and so applied to ranches of different sizes. But in making the estimates of lamb production per head to which these prices are applied, the calculation of market production per head must be on the basis of all the sheep in the bands, including a normal number of yearling sheep maintained for replacement of the death loss and of the aged and cull ewes marketed. Also the market production estimates for lamb must account for the normal requirement of ewe lambs retained or bought for replacement. For example, a ranch operates one band of 1200 ewes and 300 of these are yearlings. The 900 breeding ewes produce a lamb crop of 800 lambs, weighing an average of 70 pounds at marketing time. But 300 of these 800 are retained for breeding-ewe replacements, so 500 head are marketed. The market production of lamb is, consequently, 35,000 pounds, or 29 pounds per head for the 1200 ewes that are maintained in the operation.

This procedure for estimating the gross income of the sheep ranch is necessary to make the gross income figures comparable with per-head operating cost figures for estimating probable net income per head. It is not practical to determine the operating cost of the yearling ewes separate from that of the breeding ewes.

Production Standards. The figures for probable marketable production given in Table 8 represent the main types of western range lamb production. The sheep ranches of the plains regions and of most of the intermountain region do not have a range and feed-crop combination that favors slaughter-lamb production, and these ranches are, consequently, feeder-lamb producers. Sheep ranches of the Rocky Mountain region, part of the intermountain region, and California are able to produce either slaughter lambs or heavy lambs, a considerable percentage of which will sell in the slaughter class in a favorable range season. These figures for feeder-lamb production are based on average quality fine-wool ewes. The slaughter-lamb production estimate is for the Rocky Mountain region valley ranches and the California ranches that use crossbred ewes with mutton-breed sires, and the mixed production represents the mountain valley ranch that makes some

TABLE 8. AVERAGE ANNUAL MARKET PRODUCTION OF DIFFERENT
TYPES OF SHEEP RANCHES

Type of Ranch Operation	Average Annual Marketing per Head for All Ewes Operated on Ranch (in pounds)		
	Lamb	Cull Ewe	Fleece
Feeder-lamb production	26	6	9
Slaughter-lamb production	43	10	8
Mixed feeder- and slaughter-lamb production	35	8	9

SOURCE: These figures are based upon the production and marketing statistics of the U.S. Department of Agriculture, and upon ranch management surveys and studies of several state and federal agencies.

TABLE 9. GROSS INCOME PER HEAD OF SHEEP FOR
WESTERN SHEEP RANCHES

Type of Ranch Operation	Gross Income per Head		
	Medium High Price	Medium Price	Low Price
Feeder-lamb production	$7.20	$4.90	$2.75
Slaughter-lamb production	9.40	6.65	4.00
Mixed slaughter- and feeder-lamb production	8.25	5.75	3.25

use of the mutton breeds but has some limitations for early-lamb production.

There are individual ranches and localities whose normal production departs materially from the averages given in Table 8. The ranch that is a high producer of slaughter lambs may be able to maintain an output of fifty pounds of lamb per head of sheep operated. It is rare that the feeder-lamb producer can maintain an output in excess of thirty pounds. The figures in Table 8 for pounds of ewe marketed represent the normal sale of aged and cull ewes.

Gross Income Standards. The price figures in Table 7 applied to the production figures of Table 8 give the results in Table 9 for gross income per head of sheep operated on the ranch. The per-head income figures in Table 9 represent the ranch situation where the numbers and age composition of the breeding ewes are being maintained, and where there is no appreciable liquidation or expansion of the breeding stock capital. These figures of gross income from the sale of lambs, sheep, and wool represent the probable gross income from sheep. To this would be added any income from supplemental enterprises to arrive at the total gross income available to meet the cash operating costs of the ranch, the debt service, and the cash expenditures for family living.

CATTLE PRICES AND RANCH INCOMES

Prices. In a similar manner to the calculations for the sheep ranch, the prices received for beef cattle by the western growers can be applied to the probable production of the ranch to show what may be expected in cattle ranch incomes. There have actually been greater extremes in the prices received by the western grower during the past two decades than those shown in Table 10. The prices given in this table were selected as representing the most probable differences between high and low market price situations.

The average price of all classes of beef cattle shown in Table 10 is based upon the relative weight of marketings by western growers of the different classes of cattle during the past twenty years. However, there has been a trend during this period to market a larger proportion of the higher priced animals. This trend is caused by producing better calf crops, marketing younger animals, and, consequently, decreasing the proportion of cows in the total weight of marketings. As a result the average prices for all classes can now be figured as somewhat above the average prices given in Table 10.

Production Standards. Average annual production figures for the main types of western cattle ranches are given in Table 11. The per-head market production figures shown in this table are derived by dividing the number of animals over six months of age on the ranch on the first of January into the total weight of animals marketed. These production figures are for the average sea-

TABLE 10. HIGH, MEDIUM, AND LOW PRICES TO GROWERS FOR
WESTERN BEEF CATTLE

Market Class and Grade of Cattle	Prices per Pound to Growers		
	Medium High Price	Medium Price	Low Price
Feeder calves, "good" grade (average weight 400 pounds)	$.13	$.09	$.05½
Feeder yearlings, "good" grade (average weight 675 pounds)	.12	.08	.05
Two-year-old slaughter steers, "good" grade (average weight 900 pounds)	.11	.07½	.05
Cows, "good" grade	.09	.05½	.03½
Cows, "common" grade	.06	.04	.02½
Average of all classes	.10½	.06¾	.04½

SOURCE: This table is based upon U.S. Department of Agriculture price statistics.

TABLE 11. AVERAGE ANNUAL MARKET PRODUCTION PER HEAD OF CATTLE FOR
THE PRINCIPAL TYPES OF WESTERN CATTLE RANCHES

Ranch Types	Average Annual Marketing per Head of Cattle (in pounds)
Mountain valley and foothill ranches	325
Plains ranches	265
Intermountain and southwestern semidesert ranches	200

TABLE 12. GROSS INCOME PER HEAD OF CATTLE FOR
WESTERN CATTLE RANCHES

Ranch Types	Gross Income per Head of Cattle		
	Medium High Price	Medium Price	Low Price
Mountain valley and foothill ranches............	$34.10	$21.95	$14.60
Plains ranches	27.80	17.85	11.90
Intermountain and southwestern semidesert ranches	21.00	13.50	9.00

son and normal marketings; that is, when there is not either expansion or reduction of numbers on the ranch.

Contrasts in production shown in Table 11 for the principal types of western cattle ranches are due primarily to the differences in the resource productivity and in the management opportunities of these types of ranches. The average production figures shown in Table 11 necessarily represent a general picture of regions and types of ranches. There are, for example, some mountain valley and foothill cattle ranches with low-grade resources and low production. Also the production figures of this table represent an average of management. Differences in management result in a considerable variation of individual ranch production within these general group averages.

Gross Income Standards. We now combine the price figures in Table 10 and the production figures in Table 11 to determine the probable gross income for these types of ranches and to show the probable variations in per-head gross income due to differences in prices. Table 12 gives this result.

The gross income figures in Table 12 show the income contrast caused by the production differences of these general groups of ranch types. They also show how the gross incomes for these ranch types vary with changing prices. We should note that the production figures of Table 11 showing the production contrasts do not attempt to account for the *quality* differences that might further influence the income differential between these ranch types.

PRODUCTION AND PRICE AVERAGES FOR WESTERN STATES

A comparison by states of the average annual production yield per head of livestock and of the average prices received by growers adds something to the western picture of ranch incomes. This is true in spite of the fact that for some of the western states these averages combine the results of different natural conditions and types of ranches.

Production Averages. In Table 13 we have an analysis of the U.S. Department of Agriculture (USDA) livestock population and marketing statistics into an average annual production of lamb per head of stock sheep and of beef per head of beef cattle. These average figures, given by state for each of the eleven western states, are for the fourteen-year period from 1931 to 1944. Dairy cattle beef and the feedlot gains for beef cattle and lambs are excluded from these figures. These production figures are not, however, strictly those of the ranches. Farm beef cattle herds and farm-flock sheep are included. The results are, then, the average annual market turnoff of pounds of lamb for each head of stock sheep on ranches and farms and of pounds of beef per head of beef cattle on ranches and farms.

In the beef production figures in Table 13 for New Mexico, Arizona, and Nevada we see the influence of the semidesert cattle ranch operations of these states. We see the result of a combina-

TABLE 13. AVERAGE ANNUAL BEEF AND LAMB PRODUCTION PER HEAD OF STOCK MAINTAINED ON RANCHES AND FARMS IN THE ELEVEN WESTERN STATES FOR THE PERIOD 1931–1944

| State | Annual Average Production | |
	Beef per Head of Stock Cattle	Lamb per Head of Stock Sheep
Arizona	185	26
California	287	42
Colorado	272	38
Idaho	277	49
Montana	300	31
Nevada	214	22
New Mexico	215	17
Oregon	285	35
Utah	267	23
Washington	308	47
Wyoming	268	26

SOURCE: USDA Production Balance Sheet figures. A constant of 70 pounds per head for dairy cattle and 300 pounds per head for feedlot-fed cattle was deducted from total live-weight production to derive the live-weight production figures for cattle; and a constant of 30 pounds per head for lambs on feed was deducted from total live-weight production to derive the live-weight production figures for lambs.

tion of mountain valley, foothill, and plains ranches in the Montana figure. Probably the figure for Washington is pretty much the result of the farm beef herds. The Utah, Idaho, Colorado, and Wyoming figures all are lowered somewhat by having some aridland cattle ranching.

The lamb production figures for Idaho and Washington show distinctly the results of the early-lambing and heavy-feeding sheep ranches. For California the results of the central valley spring-lamb production are clearly depicted. A combination of different types of ranches obscures the Montana, Colorado, and Oregon results. The Arizona figure, considerably above the New Mexico figure, shows the effect of the Salt River valley and the Gila River valley irrigated-pasture lamb production. In the Nevada and Utah figures we see the results of winter maintenance on low-grade ranges.

TABLE 14. AVERAGE PRICES RECEIVED BY GROWERS FOR BEEF CATTLE
AND LAMBS IN THE ELEVEN WESTERN STATES FOR THE
PERIOD 1935–1939

State	Average Price Received by Growers (per hundredweight)	
	Beef Cattle	Lambs
Arizona	$6.35	$7.80
California	6.60	7.65
Colorado	6.80	8.35
Idaho	5.50	7.00
Montana	6.00	7.20
Nevada	6.35	7.45
New Mexico	5.80	6.75
Oregon	6.15	6.90
Utah	5.75	7.20
Wyoming	6.40	7.60

Price Averages. A comparison by states of the average prices received by growers for beef cattle and lambs shows some significant differences for the eleven western states. These differences, shown in Table 14, can be interpreted and explained to a large extent in terms of the main factors that determine the western prices.

These price averages, prepared annually by the Bureau of Agricultural Economics of the U.S. Department of Agriculture, are for all marketing. That is, range cattle and lambs, farm cattle and lamb production, and farm-fed animals enter into these averages. These averages are, however, weighted on the basis of numbers and prices of animals marketed in different classes and grades. The price averages for the 1935–39 period are somewhat below the prices shown in Tables 7 and 10 for the medium-price situation.

In the beef cattle prices for Utah, Idaho, and New Mexico, we see the influence of the quality and grade factors upon prices. The Utah and Idaho sales include a considerable proportion of farm cattle (from the irrigated districts) that are dairy breeds or mixed with the dairy breeds. New Mexico prices show the effect of a considerable proportion of thin cattle and low-quality cattle from poor ranges. The effect of proximity to the California markets may be seen in the Arizona and Nevada prices. Salt River valley feedlot finishing for the Los Angeles market also enters into the Arizona price. The Colorado and Wyoming beef cattle prices reflect the proximity to the Denver central market and to the nearby feeding operations.

Spring-lamb production and easy access to the California markets show up in the Arizona and California lamb prices. Colorado lamb prices are favorably influenced by nearness to the Denver central market. The Greeley and Fort Collins lamb-feeding operations also affect the average lamb price for Colorado.

The Effect of Market Grade upon Prices. In the usual market situation, the price spread between grades that is most important to the western cattle grower is the differential between the "good" grade and the "medium" and "common" grades. This applies more to the eastward market movement than to the western marketings. It is consequently of special importance to attain the grade of "good" with the maximum possible percentage of the marketings, through the management of livestock, grazing lands, and feed crops.

This quality factor is important in grades and prices for western range lambs, but in the general picture the attainable level of "good" grade within market classes comes nearer being realized by range lamb production than by range cattle production. It is not possible to see the distinct and significant price spread between the "good" grade and the next lower grades for the market classes of western lambs comparable with that for the market classes of beef cattle produced by the western growers.

CHAPTER VI

Business Management

Successful operation of the western stock ranch requires good financial management. Usually a highly specialized producer, seldom very diversified, the stock ranch must have good financial plans to cope with fluctuating prices and incomes. The fact that the annual income is highly seasonal adds to this need for good financial planning. Important, too, is the fact that capital investment requirements for the stock ranch run higher in ratio to the gross income than such ratio for most other kinds of agricultural production.

Compensating for these problems of management, the specialized nature of the stock ranch facilitates the use of operating cost standards and income standards as guides for the preparation of budgets and financial plans. A good set of information has become available during the past twenty years concerning the requirements and the attainable standards for ranch operating costs, gross income, net income, and capital investment. Materials presented in this chapter regarding such standards are given for the several main types of ranches and are expressed on a per head of livestock basis. This facilitates application of these standards to ranches of different size, in terms of the number of livestock operated.

OPERATING COST STANDARDS

Uses of Cost Standards. The ranch operating cost standards developed from the records of western ranches show the costs that have in the past prevailed under different price situations. Such operating cost averages or standards are given in Tables 15 and 16. These standards afford the basis, first, for analyzing the operating costs of the current year, to see how the costs of the indi-

TABLE 15. OPERATING COST STANDARDS PER EWE FOR WESTERN SHEEP RANCHES

	Annual Costs per Ewe								
Item	Early-Lamb Production (shed lambing)			Feeder-Lamb Production (range lambing)			Texas Fenced-Paddock Sheep Grazing		
	Medium High	Medium	Low	Medium High	Medium	Low	Medium High	Medium	Low
Hired labor	$2.25	$1.65	$1.15	$1.65	$1.25	$.85	$.50	$.40	$.30
Provisions and camp supplies	.65	.50	.35	.55	.35	.25	.25	.20	.15
Shearing	.30	.20	.15	.25	.15	.10	.30	.20	.15
Feed and salt	1.10	.75	.50	.50	.30	.20	.50	.35	.20
Leases and grazing fees	.75	.60	.45	.50	.40	.30	.80	.65	.50
Buck purchase	.30	.20	.15	.25	.20	.10	.25	.20	.15
Auto and truck expense	.30	.20	.15	.25	.15	.10	.20	.15	.10
Taxes (land and sheep)	.55	.40	.25	.45	.35	.25	.30	.20	.15
General ranch expense	.40	.30	.20	.35	.30	.20	.40	.30	.20
Depreciation on equipment	.30	.30	.30	.20	.20	.20	.30	.30	.30
Total operating cost per head*	$6.90	$5.10	$3.65	$4.95	$3.65	$2.55	$3.80	$2.95	$2.20

*During the period of high agricultural prices from 1944 to 1948 the annual cash operating cost per ewe rose to approximately $13 for the shed-lambing ranches, $8 for the range-lambing operations, and $6 for the Texas ranches.

TABLE 16. OPERATING COST STANDARDS PER HEAD OF STOCK CATTLE FOR WESTERN CATTLE RANCHES

Annual Costs per Head

Item	Mountain Valley and Foothill Ranches			Plains Ranches			Intermountain and Southwestern Semidesert Ranches		
	Medium High	Medium	Low	Medium High	Medium	Low	Medium High	Medium	Low
Hired labor	$5.85	$3.75	$2.75	$3.60	$2.65	$1.85	$2.75	$1.85	$1.35
Provisions and camp supplies	2.25	1.50	1.10	1.60	1.15	.75	1.15	.75	.45
Feed and salt	2.50	1.65	1.25	2.75	1.85	1.25	1.85	1.35	.80
Leases and grazing fees	1.75	1.20	.85	2.25	1.65	1.20	1.25	.85	.50
Bull purchase	1.85	1.25	.85	1.85	1.25	.85	1.20	.90	.60
Auto and truck expense	1.15	.85	.65	.85	.55	.40	.65	.50	.35
Taxes	1.75	1.35	.90	1.45	1.15	.85	1.20	.95	.75
General ranch expense	2.45	1.65	1.35	1.90	1.25	.95	1.35	.75	.55
Depreciation on equipment	1.15	1.15	1.15	.75	.75	.75	.65	.65	.65
Total operating cost per head*	$20.70	$14.35	$10.85	$17.00	$12.25	$8.85	$12.05	$8.55	$6.00

*In 1948 the total annual operating costs per head had risen, with the recent all-time high beef cattle prices, to approximately $35 for the mountain valley cattle ranches, $28 for the plains ranches, and $20 for the semidesert ranches.

vidual ranch vary from these standards and the reasons or lack of reasons therefor; and, second, for making the ranch budget and financial plan for the year ahead, by furnishing probable operating costs.

The operating cost standards given in Tables 15 and 16 for the different types of ranches are derived from ranching survey studies, and they represent a cross section of the operating records of many hundreds of ranches of each type. The individual ranch of a certain type may justifiably depart considerably *in specific cost items* from the cost standards for its type. Such a departure may be caused, for example, by variations in the extent to which crop feed is raised or purchased. But the individual ranch should not as a rule depart a great deal from the total per-head operating costs given in the tables for its type. The medium high, medium, and low price situations used in developing these cost standards are those shown in Tables 7 and 10.

It should be emphasized in applying the operating cost standards of Tables 15 and 16 to an analysis of the operations of the individual ranch that these standards apply to ranches of sufficient size to be classed as a commercial operation. For an operation of such size the operator's work is primarily that of management, and the labor of the operator and his family is not a major factor in the total labor requirements of the ranch. These standards apply reasonably well to the ranches operating at least 150 head of cattle or 750 ewes, but they are particularly applicable to the ranches that operate 300 head or more of cattle, or 1500 head or more of ewes.

These operating cost standards do not include personal expenditures of the operator and his family. That is, the commissary account is for hired labor only, and the auto expense is for the ranch business only; the figures were determined by an estimated allocation of this expense item between ranch operation and personal use. The only item in these operating expenses that might be construed as a personal item is the inclusion of property taxes on ranch dwellings in the tax cost. The item of taxes includes all the property tax expense for the ranch; that is, real estate taxes and personal property taxes on livestock and equipment.

All the cost items in Tables 15 and 16 are cash costs with the exception of depreciation on equipment. This item is a cash cost indirectly, since the depreciable items are first capitalized (generally on the basis of their cash cost), and the capital value of the items is charged off as an annual cost during the life of the equipment. No depreciation cost for fences is included in the depreciation expense item. Fence repairs and the cost of necessary annual rebuilding of fences are included in the item "general ranch expense."

The feed and salt cost item is the average annual cost for the purchase of grain, concentrate feed supplements, minerals, and hay. As was pointed out previously, the ranch that buys a large part of its hay should have a lower cost for hired labor. The leases and grazing fees item is the annual cash cost for land leases and public land permits. A ranch that leases a large part of the grazing land used will usually have a larger expenditure for this item and a smaller cost for land taxes. General ranch expense includes livestock supplies (dip, vaccine, and the like), equipment repairs, property insurance, and the large number of small items that do not fit any of the main cost accounts.

Sheep Operating Costs. The operating cost standards for sheep ranches given in Table 15 reflect the contrast in operating methods of different types of sheep ranches. Early-lambing operations have the highest costs, particularly for labor, feed, and range. An exception to this in range costs is noted for the Texas ranches which are range-lambers with a high rangeland cost, because they operate entirely on fenced range. Operating cost standards given in Table 15 for range lambing, other than for the Texas operations, apply particularly to the sheep ranches of the northern and central plains and the intermountain region. A further breakdown of this for the range-lambing operations of the intermountain region would show that they have a somewhat lower feed and range cost than these averages, owing to the use of winter range at low cost on public lands. Because of their yearlong grazing on fenced range and their lower wage rates for ranch labor, the Texas operations are unique in their low labor cost.

It will be noted that no cost item for depreciation on breeding animals is shown in the tables, since the usual practice is to retain breeding herd replacements from the animals raised by the ranch. When the breeding animal replacements are purchased (often the practice of sheep ranches that use mutton-breed sires with fine-wool ewes), the usual accounting practice is to consider the purchase cost of the ewe replacements in the calculation of gross revenue, rather than to regard the purchase of replacements as an annual operating cost. That is, in accounting for gross revenue on an inventory basis, the accounting procedure is: (receipts from sales + the value of the livestock inventory at the close of the year) — (the value of the livestock inventory at the start of the year + purchases) = gross revenue.

Since the general practice of the ranch is to purchase the sires required for replacement, an average annual cost for sire purchase is shown in Tables 15 and 16. This cost item represents the usual practice of using bucks through four to five seasons, the bulls through three to four seasons. Death loss of sires reduces the average period of use that is actually obtained from the sires.

Annual operating costs as shown in the tables for medium high, medium, and low prices are what may be expected when these prices have prevailed for two or three years — that is, long enough for the cost rates to become adjusted to the price and income situation. No variation is indicated for the cost item depreciation on equipment, since the depreciable items — sheds, machinery, tools, and the like — usually have a life extending over a period of five or ten years or longer. The item of shearing in Table 15 includes the commissary cost for the shearing crew, the bags, twine, and so on. The per-head operating costs for sheep in Table 15 are based on all sheep at the start of year, including the normal number of ewe lambs for replacement which are carried until they are of breeding age.

Cattle Operating Costs. Differences in operating costs as shown in Table 16 for the main types of cattle ranches result principally from the different winter feed requirements as they affect labor costs and the purchase of feed, and from the differences in the productivity of the resource as reflected in land costs. Cattle ranches of the plains have the highest cost for purchased feed, primarily because of the purchase of concentrate feeds as a range and hay supplement. The plains ranches also have the highest cost for leases and grazing fees, but this may be explained by the fact that they lease a large amount of privately owned rangeland. However, the mountain valley and foothill ranches have a higher *total* cost for land than the plains ranches, because of the higher percentage of deeded land used by the mountain valley and foothill ranches. This is reflected in their higher land taxes and the higher interest paid on land investment. The lower level of costs for the intermountain and southwestern semidesert ranches shows the adaptation that these ranches have of necessity made to the lower productivity of their resources.

It should be noted here that the cattle ranch operating cost standards of Table 16 do not represent regional differences but rather types of ranches that can be generally characterized by regions. There are many mountain valley and foothill ranches in the intermountain regions that are comparable to those of the Rocky Mountain region; there are cattle ranches in the southern plains that are essentially semidesert ranches in their natural resources and production.

The bull purchase item in Table 16 is the cost of the average annual expenditure for replacements. Mountain valley and foothill ranches use a somewhat larger number of bulls per hundred head of females than do the plains ranches, but since the plains ranches operate more nearly on a breeding herd basis they have about the same number of bulls per hundred head of all cattle as the mountain valley and foothill ranches do. That is, in arriving at the bull purchase cost figure given in Table 16 the cost of bull purchase is prorated to all of the cattle, not just to the females of breeding age. As a general average for recent years the number of females of breeding age has been 60 percent of the total number for the plains ranches, 50 percent for mountain valley and foothill ranches, and 40 percent for the intermountain and southwestern semidesert ranches. The annual maintenance cost of the bulls is included in the operating cost averages per head for the total number of cattle operated.

Interest as a Part of Land Costs. The operating cost standards shown in Tables 15 and 16 do not include interest as a feature of the land costs, neither interest paid on land obligation nor interest imputed to the value of the proprietorship value in the ranch. Too much variation prevails in the amount of interest paid on land and livestock obligations to consider either of these items as a cost or to set up any standards for these items.

Where a ranch operates entirely on deeded lands it will have a higher "cost" for land interest — in either of the senses mentioned

above — than the average for the ranches from which the standards of Tables 15 and 16 were taken, because these ranches were
leasing one third to two thirds of their rangelands. The ranch that
operates entirely on deeded land can, consequently, expect to have
a tax cost about a fourth higher than those shown in the tables.

NET INCOME STANDARDS

Totals for the ranch operating costs, as given in Tables 15 and
16, subtracted from the gross income figures shown in Chapter V
give us our standards for net income. Net income as thus expressed is the annual amount available to carry the debt service,
to remunerate the operator for his work, and to pay the interest
return upon the operator's equity in the ranch property and the
livestock.

How Net Income Is Determined. The gross income figures used
in Tables 17 and 18 are taken from Tables 9 and 12. These gross
income figures do not include the items of other income from supplemental enterprises and by-product sources which are important
for some ranches and relatively unimportant for others. Most
sheep ranches have a small amount of income, other than sheep,
lamb, and wool sales, from such sources as the sale of pelts or
cattle sales from a small beef herd. A few cattle ranches maintain
a horse enterprise that produces some income. Sale of alfalfa seed
or grass seed is sometimes an important supplemental income, but
this is generally an unpredictable and "windfall" type of income.

Where items of other income are of sufficient importance and
regularity they should be accounted for in the budget and financial plans of the ranch. The operating cost averages shown in
Tables 15 and 16 cover the minor additional costs of producing
the items of other income realized on about half of the ranches

TABLE 17. NET INCOME STANDARDS FOR WESTERN SHEEP RANCHES

| Item | Sheep Ranch Standards | | | | | |
| | Early-Lamb Production (shed lambing) | | | Feeder-Lamb Production (range lambing) | | |
	Medium High	Medium	Low	Medium High	Medium	Low
Gross income per head	$9.40	$6.65	$4.00	$7.20	$4.90	$2.75
Operating cost per head	6.90	5.10	3.65	4.95	3.65	2.55
Net income per head........	$2.50	$1.55	$.35	$2.25	$1.25	$.20

TABLE 18. NET INCOME STANDARDS FOR WESTERN CATTLE RANCHES

| | Cattle Ranch Standards | | | | | | | | |
| Item | Mountain Valley and Foothill Ranches | | | Plains Ranches | | | Intermountain and Semidesert Ranches | | |
	Medium High	Medium	Low	Medium High	Medium	Low	Medium High	Medium	Low
Gross income per head	$34.10	$21.95	$14.60	$27.80	$17.85	$11.90	$21.00	$13.50	$9.00
Operating cost per head	20.70	14.35	10.85	17.05	12.25	8.85	12.05	8.55	6.00
Net income per head	$13.40	$7.60	$3.75	$10.75	$5.60	$3.05	$8.95	$4.95	$3.00

from which the cost information was taken. However, since the average of the figures for other income was not too typical or significant, no calculation for other income is included in Tables 17 and 18. These net income figures are the gross income minus the ranch operating costs shown in Tables 15 and 16.

These net income figures represent the amount of cash income that would normally be available for debt service and personal withdrawal. In any one year this amount may be modified materially by supplemental income sources of the ranch.

No separate calculation is shown in Table 17 of the net income standards for the Texas fenced-paddock sheep operations. Indications are that the lamb and wool production and the gross income of these Texas operations does not differ materially from that of the range-lambing and feeder-lamb operations of the northern and central plains regions. Because of their lower costs these Texas operations have, consequently, a very favorable per-head net income. This accounts for the high capitalization of rangelands and improvements carried by these Texas ranches.

In Table 18 are shown the standards for gross income, operating cost, and net income for the three main types of cattle ranches. These net income figures are derived by the same procedures described for Table 17.

Taken together, the operating cost and income standards given in Tables 17 and 18 provide the basis for preparing the annual financial plan of the ranch. This plan can be, and should be, prepared in systematic form to constitute the financial budget that is needed for an orderly financial program and for financial control over the current operations of the year.

BUDGETS AND FINANCIAL PLANS FOR THE RANCH

A budget gives a systematic statement of the financial plan, the expectations and probabilities for the year. The financial plan, as expressed in the budget, has two main management objectives: (1) maintaining control over the operating costs by periodic (monthly or quarterly) comparison of the actual operating costs with the budget figures, and (2) attaining a balance through the year of cash receipts and payments, including the credit arrangements.

Operating Cost Estimates. In order to serve the first of the two purposes stated above, the budgetary forecasts of the expenditures for the coming year's operations should be set up to show totals for each month and for each quarterly period of the year. The operating cost items in the budget estimates for expenditures should be comparable with the operating cost accounts. This latter requirement of a budget is illustrated by the budget form (Form A) for estimating the expenditures of a sheep ranch. The items for estimating costs of a sheep ranch in this form are the same as the sheep ranch operating cost accounts used in Table 15. An easy comparison can then be made at the close of each quarterly period between the budget estimates and the accounting of actual costs. While the items of operating cost for making the estimates in Form A parallel the operating cost accounts of Table 15, the items are shown in greater detail in the form to facilitate the making of the budget estimates.

Income Estimates. The budget estimate of expected income should also be prepared by months and summed up by quarterly periods for the year. This facilitates comparison with the budget estimate of cash expenditures and the making of an orderly plan of finance for the year's operations. The income items for making the budgetary estimates used in Form B are also for a sheep ranch, in keeping with the illustrative material used in Form A.

The Year's Financial Plan. The estimates of cash expenditures and of cash receipts are together an expression of the financial plan for the year. These budget estimates, as illustrated in Forms A and B, *do not* constitute an estimate of the *net income* from the year's operations, because estimates of cash receipts and expenditures include borrowing and repayments and personal withdrawals by the operator and do not account for changes in the inventory items of the ranch. The increase or decrease in cash or bank balances is *not* the equivalent of the net income from the year's operations. The budget sets up the financial plan of receipts and expenditures for the year, and it gives some standards for comparison with the financial accounts, particularly the operating cost accounts, as the year progresses.

ACCOUNTS AND RECORDS

Primarily, the purpose of a system of accounts and records should be analysis, planning, and control, rather than historical review. Financial accounts are not in themselves sufficient for this purpose; they must be accompanied by certain records kept in terms of physical quantities. The term *records*, as used here, refers to the physical operating data of the ranch; the term *accounts* refers to the financial data. Financial accounts cannot be interpreted or analyzed without adequate physical records.

Types of Accounts. Four kinds or classes of accounts are used in any complete accounting plan: (1) asset accounts, (2) liability and proprietorship accounts, (3) income accounts, and (4) operating cost accounts. The asset accounts and the liability and proprietorship accounts constitute the starting point in the year's financial accounting. These accounts make up the balance sheet of the business, what is sometimes called the "property statement." Balance sheet accounts, as adapted to the stock ranch, are illustrated by Form C.

The accounts illustrated in Form C are for the unincorporated business. For an incorporated business the proprietorship accounts would consist of the corporate shares and any undivided surplus from past profits. The owner's investment account given in Form C is charged at the close of the accounting year with the amount of the operator's personal withdrawals through the year as shown by the operator's personal account, and it is credited with the net profits of the year's business as shown by the summation of the operating expenses and income through the profit and loss account.

Asset and liability accounts, as shown in Form C, are classified into current and fixed asset accounts, and current and fixed liability accounts. A significant business management relationship prevails between the current assets and the current liabilities and between the fixed assets and the fixed liabilities. Current assets constitute the working capital items. As a general rule-of-thumb guide the total normal value of the current assets should be at least three times the total current liabilities if the ranch business is to have sufficient working capital. If the ratio is much below this, it is likely to be difficult to work out an orderly program of current working capital loans and repayments (an average turnover period for the value of the current assets of a ranch is about three years for the cattle ranch and two years for the sheep ranch). The relationship between the value of fixed assets and the fixed liabilities is generally considered satisfactory when the total normal value of the fixed assets is twice that of the fixed liabilities.

Operating cost accounts adapted to the needs of the stock ranch are illustrated by the accounts shown in Tables 15 and 16. An income account should be set up to show the amounts of the purchases and the amounts received in sales from each of the respective items that are important income sources — cattle, sheep, wool, feed, and so on — with a miscellaneous income account to accumulate the total for small and relatively unimportant sources of income.

Through the year the operating cost accounts and the income accounts accumulate the information showing the results of the year's business transactions — total operating cost, gross income, and net income. The cost and income accounts are summed and closed into a summarizing account (profit and loss account) at the end of the year, and the net profit or loss shown is carried to the owner's investment account to determine the amount of the owner's investment for the closing balance sheet. Thus the balance sheet constitutes the starting point and the closing point in the accounts of the year's operations.

Accounting for Gross and Net Income. Cash receipts from sales do not as a rule reflect the true gross income from the year's operation, because of livestock purchases and inventory changes and the possible deferment of cash receipts on a sale contract beyond the close of the accounting year. It is desirable to determine separately the gross income from each major enterprise of the ranch if there is more than one enterprise, such as cattle, sheep, hay, livestock fattening.

There is not, however, much point in making gross income calculations for the hay, grain, and other feed crops when the principal outlet for such crops is the livestock maintained on the ranch. Gross income for each enterprise is determined by adding the sales (including sales for which cash payment has not yet been received) and the value of the closing inventory, and then subtracting from this total the sum of the purchases and the value of the inventory at the start of the year.* This will give the *gross* reve-

* It is generally desirable to use a constant value figure from year to year in arriving at the values of the livestock inventories for gross revenue determination. The use of current market prices for inventory valuation is likely to result in showing "paper" profits or losses which may not be realized in the subsequent year or years. The constant value figure for livestock inventory valuation may be a long-time average market price that is considered a good normal figure, or it may be the normal livestock production cost figure for the ranch if such costs are known.

nue from each enterprise. A determination of the *net* income for each of the several enterprises that may be operated by the ranch requires enterprise cost accounting for allocating the proportional share of the operating costs to each enterprise. The methods for such enterprise cost allocation are presented later in this chapter.

For the ranch that does not undertake enterprise cost accounting to determine the *net* income from each enterprise, the method for determining the net income (from all enterprises) is to subtract the total of the operating cost items from the gross income. Form D illustrates this, and also a method for determining the gross income of each enterprise.

In the preparation of the income statement, the cash operating costs should be adjusted at the close of the accounting year for any deferred or prepaid items. For example, a lease rental may be paid for three years in advance, as shown by the lease records, and only one third of this cost should be accounted for in the current year's operation. Conversely, there may be an unpaid open book account for commissary supplies at the close of the accounting year that should be added to the cash expenditures in this account, if the true picture of the year's operating cost is to be shown. The cost item for depreciation on equipment is calculated from the equipment inventories, which should list the principal items of the equipment. These inventories should show for each important item of equipment and for each improvement the original cost or appraised value for the depreciation value base, the expected life of each item, and the annual depreciation charge.

Balanced Accounts. For the ranch operator wishing to keep balanced accounts a double-entry journal and a ledger are the two primary accounting books. The journal is the daily record of transactions, in which debit and credit entries are made for all receipts and for payments by both cash and check. The ledger contains the accounts — the balance sheet accounts and the income and operating cost accounts. At the start of the year the only accounts in the ledger that contain balances are the balance sheet accounts. Through the year the entries in the journal are posted to the ledger accounts, and a periodic balance is made of the ledger accounts.

At the close of the year the ledger accounts are balanced as the starting point for preparing the financial statements, that is, the income statement and the balance sheet. For closing the ledger and preparing the financial statement it is customary first to prepare a work sheet on multicolumn paper. In the first two columns of this work sheet the balances in the ledger accounts at the close of the year are written (see Form E). Next, the *adjusting* entries are made to bring in the inventories for the end of the year and the items of prepaid and deferred income and expense. The *closing* entries are then made to transfer the sales, purchases, opening inventories, and operating costs to the profit and loss account and to transfer the net profit or loss to the proprietorship account.

These adjusting and closing entries are illustrated in the second set of columns in Form E. Each of these entries is keyed by a number to show the corresponding debit or credit for each such entry. Adjusting entries are those which bring new figures into the accounts. Closing entries are those which, in summarizing, transfer a balance from one account to another. Number 7 is a closing entry.

The results of the account balances in the first set of columns plus the adjusting and closing entries in the second set of columns are shown in the third and fourth sets of columns under the headings "Profit and Loss" and "Balance Sheet." These columns are the basis for the preparation of the two financial statements — the profit and loss statement and the balance sheet — at the end of the year. The adjusting and closing entries as shown in the work sheet are then recorded in the journal and posted to the ledger accounts to close the accounts and set up the new ledger balances of the balance sheet accounts for the start of the year. This device, illustrated in Form E, affords a convenient basis for the preparation of the financial statements of the ranch even though the plan of accounting is not one of balanced accounts.

Accounting without Using Balanced Accounts. It is not necessary for ranches of small and medium size to have a system of double-entry bookkeeping in order to attain reasonable proficiency in their financial accounting. (The principal merit of double entry is that it promotes completeness and accuracy in the accounting.) The best alternative to double-entry and balanced ledger accounts is to use (1) a multicolumn journal for recording the cash expenditures, (2) a similar journal for cash receipts, and (3) adequate inventory records.

We will develop this alternative, rather briefly. The accounting forms given in this book are of the author's design, and they have been widely used by western ranches. The illustrations used are from actual ranch accounts. Form F illustrates the two multi-column cash journals, Forms G to L illustrate the inventory records. In the cash payments journal all payments by cash or check are recorded, the total amount is shown in column 5, and this amount is distributed to the column or columns representing the proper account or accounts. Each receipt is recorded in the cash receipts journal. The total amount is shown in column 4 of this journal; that amount is then entered under the proper column heading or headings. The account columns of these two journals are summed at the end of the year, the journals thus serving the purposes of both the daily record and the ledger accounts for all cash transactions. Memorandums should be made of noncash transactions of any consequence.

These two special-column cash journals must be accompanied by a rather complete "inventory record" of all asset and liability items at the beginning of the year and at the close of the year. These inventory records give the necessary information for the valuation of the livestock in accounting for the gross income (that is, on an "accrual" basis rather than on a cash basis). The inventory records also give the basis for the valuation of equipment and any depreciable ranch improvements in the accounting for depreciation, and for adjusting the income accounts and the operating accounts to show the effects of any prepayments or deferments of cash expenditures or of cash receipts.

The year-end column totals of the special-column journals, such as those illustrated in Form F, the inventory records such as those illustrated in Forms G to L, and the use of the work sheet illustrated in Form E afford a reasonably good basis for financial accounting without the use of a ledger and balanced accounts. Such a plan of accounting is adequate for all but the large ranches.

The inventory records (see Forms G to L) that should be made at the start and close of the year for financial accounting, whether balanced accounts are used or not, are the following: (1) livestock numbers and values; (2) feed supplies and values; (3) building and equipment values and depreciation rates; (4) payable items such as notes, contracts, open accounts, wages, leases, and taxes; (5) receivable items such as notes, open accounts, prepaid leases and wages; and (6) lands and range use.

Whenever possible the inventory records should be so arranged as to serve the needs of financial accounting and also to supply the needed physical information for management analysis. The livestock inventory record should show the periodic livestock counts; the death losses; the calf and lamb crops docked, branded, and raised; the sale weights; and the results of the different ewe bands in lamb production, lamb weights, losses, marketing, dates, and the like. Along with this record of livestock performance, the rangeland and feed-crop inventories should show a record of the use of range and feeds — the number of animals and dates on and off each range; condition of the range; feeding dates and rates for hay and other feeds; and the use of range supplements. Often these types of management information are kept in diary or memorandum form, but they can be formalized and systematized in the inventory records, as shown in Forms G through L.

Payroll and Labor Time Record. If the stock ranch is to attempt a cost analysis of such things as equipment operation costs, job costs, hay production costs, or production costs of different classes of livestock, it is necessary to keep certain physical information on the labor time and the equipment time used in the various operations and enterprises. One means for doing this on an estimate basis and without burdensome detail is to keep a combined labor payroll and time distribution record similar to the one illustrated in Form M.

This record form serves a dual purpose — it gives a financial record for each employee and it provides a basis for estimating at the end of the month the workdays going to various operations and enterprises. This record of employee time through the month plus the operator's knowledge of the type of work performed by each employee provides the basis for an estimate on this form at the close of each month of the amount of labor and equipment time devoted to each enterprise or to specific operations or jobs.

In closing the subject of accounts and records for the management of the stock ranch, it should be pointed out that the natural operating year of the stock ranch is usually something other than the calendar year, since it begins after the annual marketings of livestock. The use of accounts and records for management may be facilitated by the use of the natural operating year as the starting and closing point for the accounts and records.

Types of Cost Analysis. The following are the types of cost analysis most useful in the management of the stock ranch:

1. Livestock production costs for different age classes of animals, as a basis for choosing the class of animal to market.

2. Hay and other feed-crop production costs, as a basis for choosing between the production and purchase of feeds, or for determining the profit opportunities in the production of feeds for sale.

3. Job costs — such as mowing, raking, and stacking hay — as a basis for comparing the efficiency of different methods, for deciding whether to hire certain jobs done on a contract basis, or for determining the rates at which job custom work can be done for others.

4. Equipment time operating costs in relation to work output as a basis for choice in the use of different types of equipment — for example, horse-operated equipment and mechanical power equipment.

The first two of these types of cost analysis are often referred to as "enterprise cost accounting," since the purpose is to allocate the costs of the ranch to the different kinds of production in which the ranch may be engaged. Usually the objective of this analysis is to determine which kinds of production are the most profitable. There are limitations on this type of cost analysis — the stock ranch may not have alternatives in production enterprises, or two or more production enterprises may be highly interdependent in the use of the resources of the ranch and in effecting a good distribution of labor time through the year.

Methods of Enterprise Cost Analysis. One method of analysis for determining an enterprise cost begins by separating the operation into the main jobs of the production. For the stock ranch this method applies especially to the feed crops, since the feed-crop enterprises can usually be analyzed into jobs. For example, the production of alfalfa hay can be analyzed into the following operations: (1) irrigating, (2) mowing, (3) raking, (4) stacking, and (5) periodic renovation of the hay stands. For calculating the cost of each of these jobs, it is first necessary to obtain information on the equipment and labor time requirement of the job from a time record such as that illustrated in Form M. Next, the cost rates for the labor and equipment must be found.

When the labor and equipment time requirement for a job is determined, the labor cost is then calculated from the wage rates, and the cost of the equipment time from the financial records that show the repair costs, depreciation costs, and fuel and other operating costs. (Where horse-operated equipment is used, the maintenance cost of the horses should be calculated in terms of the value of the livestock production capacity of the ranch which the horses displace.) After both the time requirements and the cost rates of the job, for labor and equipment, are determined, the cost of the job is calculated by multiplying the labor and the equipment time requirements by the labor and equipment cost rates. Finally, the total product cost for the enterprise is found by summing the job costs for the jobs that go into the enterprise.

Another method of enterprise cost determination often used does not require the analysis of the enterprise costs by job costs. Though this method is less exact than the other it is useful in determining the approximate cost for enterprises that cannot easily be analyzed into distinct operations. An example of this is the determination of the separate production costs for the different age classes of cattle produced and marketed by the ranch, or of production costs for cattle and sheep produced on the same ranch.

In this method the first step is to determine the sum of those operating costs that can be directly allocated to each of the enterprises. Then the total of the operating costs for all the enterprises of the ranch are prorated or distributed to each enterprise in ratio to the known directly allocable costs. Sometimes this distribution of the total cost to the enterprises is made in ratio to the one most important item of the directly allocable costs, such as the direct labor cost for each enterprise.

As an illustration of this method, a cattle ranch has a program of marketing some calves, some yearlings, and some two-year-olds. Each of these types of marketings may be considered a production enterprise. This ranch has been able to determine from the ranch records and accounts the winter feed production costs and the amount of feed used by the cow herd, the calves, and the long yearlings. The feed cost is, consequently, directly allocable to the different market age classes. The cost of the rangeland in leases, taxes, interest, and maintenance of improvements is determined from the accounts and allocated to the calves, yearlings, and two-year-olds on the basis that cows equal one animal unit, yearlings 65 percent of an animal unit, and two-year-olds 85 percent of an animal unit. These direct costs for feed and range are summed for each class of animal and the total operating cost for the year is prorated to the calves, yearlings, and two-year-olds in ratio to these directly allocable costs. The use of this method requires judgment in determining the reasonableness of the basis used for prorating costs that are not directly allocable to the different enterprises.

Some Results of Enterprise Cost Analysis. The use of this second method of enterprise cost analysis in a study of the operations of a considerable number of foothill cattle ranches of Montana and Wyoming gave the figures shown in Table 19 on production costs per head for different age animals.

It should be emphasized that the per-head production cost figures shown in Table 19 are averages and consequently represent an average of several different production opportunities and degrees of management efficiency. An annual depreciation charge of $2 per breeding cow was used in calculating the calf production cost. This charge may appear to be low, but it is based on the fact that a percentage of the cows are sold as dry and fat cows, with little depreciation cost. An interest charge of $5 per cow (5 percent on a $100 value in the cow and the ranch) and an operator

TABLE 19. PRODUCTION COSTS PER HEAD FOR CATTLE OF DIFFERENT AGES

Age Groups	Production Costs per Head		
	Low	Medium	Medium High
Calves			
Excluding interest and operator wage.........	$17–22	$26–31	$35–40
Including interest and operator wage..........	25–30	34–39	43–48
Long yearlings			
Excluding interest and operator wage..........	28–33	43–48	58–63
Including interest and operator wage..........	34–39	49–54	64–69
Long two-year-old steers or cows			
Excluding interest and operator wage..........	39–44	58–63	80–85
Including interest and operator wage..........	46–51	65–70	87–92

wage charge of $3 per cow was added to the ranch operating cost per cow to determine the per-head calf production cost including interest and operator wage. Three fourths of this interest and operator wage charge of $8 per cow ($6) was added to the annual ranch operating costs for the long yearling, and seven eighths ($7) to the ranch operating cost for the long two-year-olds. The average calf crop for the ranches from which these cost figures were derived was about 75 percent for the cows in the breeding herds.

Production Efficiency Cost Analysis. The objective of the third and fourth types of cost analysis described earlier is to determine relative efficiency in the performance of jobs and to aid in the choice between operating methods, rather than the choice of the kinds of enterprises which the ranch might operate. Sometimes such job cost analysis is used to determine the charge for doing custom work. Though this kind of cost analysis for the stock ranch applies principally to the production of hay and other feed crops, it may also apply to the management of ranges and livestock. For illustration, the relative cost of substituting fencing for labor in the control of livestock on the range might be a subject for this kind of cost analysis.

Job cost analysis procedure has already been described for the making of enterprise cost estimates where several distinct jobs can be recognized in the costs of an enterprise, and it need not be repeated here. As an illustration of job cost analysis, a ranch using a certain type of power equipment for mowing hay wishes to know the per-acre cost of this job for this type of equipment. The records show that the average time performance of this equipment is three fourths of an hour per acre for mowing. The labor cost for operating the equipment is $.75 an hour. The operating cost for the equipment for fuel, servicing, repairs, depreciation, taxes, and interest is calculated from the records and accounts as being approximately $1.25 per hour of operation for the normal season. The cost to this ranch for the mowing job with this equipment is, consequently, $1.50 an acre (¾ of $1.25 plus $.75).

In any comparison of the costs of different methods of performing a job, it is essential that the comparison be made on a basis of unit of work output. For example, mechanized equipment may be more costly per hour or for certain jobs, but less costly per unit of work output, that is, per ton of hay or grain.

Cost analysis for comparison of the relative efficiency of methods must be tempered with a consideration of management factors other than the cash operating costs. One method may be more costly per unit of work output than another but may shorten the total time required for an operation sufficiently so that the timeliness of performing the operation is measurably increased. This may reduce peak loads on labor and equipment time, reduce loss from spoilage, or avoid time conflicts between the different operations to such a degree that the more costly operating method is also the more profitable to the ranch as the management unit.

Finally, it should be pointed out that there is seldom, if ever, justification, even on the large ranch, for maintaining a cost accounting "system" as a continuous and integral part of a plan of records and accounts for management. The cost analysis should attempt to develop a reasonably good estimate from the operating records and accounts of the ranch, for whatever special features of costs appear to be timely and useful to management.

INVESTMENT REQUIREMENTS AND INVESTMENT FINANCING

Capital Requirements. The capital requirements and the long-time financial planning of the ranch, as distinguished from the annual operating budget and financial plan, are another phase of ranch business management for which some standards and guiding principles have been developed.

First, let us take a look at the physical setup and the capital investments of an individual ranch, to show in some detail the items that enter into the capital requirements of the stock ranch. The illustration in Form N is a specific case selected as fairly typical of the foothill cattle ranches of the Rocky Mountain region. The values used are considered, by past standards, to approximate long-time average or normal values.

The ranch used for Form N operates entirely on deeded land, but the land investment compares closely with the average for foothill ranches which generally use some leased range. The land values for this ranch include the costs of fencing and water development. The equipment is valued at cost less depreciation.

A comparison of the investment value of the ranch illustrated with the standards for foothill cattle ranches given in Table 20 shows that this ranch conforms rather closely with these stand-

TABLE 20. AVERAGE INVESTMENTS OF FOOTHILL CATTLE RANCHES AND MOUNTAIN
VALLEY SHEEP RANCHES, 1935–1940

Item	Average Investment per Head		Percent of Total Investment	
	Cattle Ranches	Sheep Ranches	Cattle Ranches	Sheep Ranches
Livestock	$50	$7.50	40	29
Grazing land	40	6.75	31	25
Hayland and cropland..........	15	3.25	12	13
Buildings	8	3.50	6	13
Equipment	5	2.50	4	10
Work stock	3	.75	2	3
Feed and supplies	7	1.75	5	7
Total	$128	$26.00	100	100

ards, except that it has a high investment in equipment. Form N
shows the ranch investment items in some detail. Table 20 gives
some additional standards for ranch investments and the distribution of these investments among the main items.

The investment standards for foothill cattle ranches given in
the table are based upon an analysis of sixty-five ranches located
in Wyoming, Colorado, and Montana. These ranches operate primarily on deeded rangeland, about 85 percent deeded as an average. Since these ranches represent one of the most productive
types of western cattle ranches, the investment standards given in
Table 20 are too high for the plains ranches and the intermountain and southwestern semidesert cattle ranches.

The investment standards shown for mountain valley sheep
ranches are the average for fifty-eight ranches located in western Montana, southeastern Idaho, and western Wyoming. These
are shed-lambing operations; most of them use national forest
summer range. Their use of public land is reflected in a somewhat
lower percentage of total investment in rangeland than that of the
foothill cattle ranches. This type of sheep ranch carries a considerably higher total investment than the range-lambing operations
of the plains and intermountain regions, primarily because of the
higher investment in buildings and equipment and in cropland by
the mountain valley sheep ranches.

Investment Financing. Referring again to Form N, the cattle
ranch used for this case illustration carries a livestock production
credit loan of $6000, approximately one third of the normal value

of the livestock, and a land credit loan of $10,000, one half of the normal value of land. The livestock loan is on an annual basis, and the land credit is on a five-year-term land mortgage. Both these loans are near the maximum which any good financial plan would attempt to carry. However, the five-year-term mortgage for the land credit is hazardous, and a fifteen- or twenty-year plan of amortized land credit would place this ranch in a much stronger financial position. Progressive land credit agencies are now offering such a financial plan for ranches. Some such agencies provide that the borrower can make additional payments in years of good income, as a prepayment reserve against years of low income. Such a prepayment reserve earns interest at the loan rate.

Good planning for long-term finance has in the past been pretty much lacking in the business management of the western stock ranch. It is only since about 1935 that the land credit agencies have begun to investigate geographic areas and types of ranches for possibilities in long-term land finance. With some exceptions, western stock ranches offer such possibilities, and the realization of them by ranch operators and lending agencies will add a good measure of stability to the management of western stock ranches.

FORM A. ESTIMATES OF CASH EXPENDITURES FOR A SHEEP RANCH

Item	January			February			March			Three-Month Total Amount
	Number or Quantity	Rate	Amount	Number or Quantity	Rate	Amount	Number or Quantity	Rate	Amount	
Herders	2	$75	$150	2	$75	$150	2	$85	$170	
Camp tenders										
Lambers										
Ranch labor	1	$60	$60	1	$60	$60	2	$75	$150	
Hay labor										
Shearing										
Tagging							2400 head	$.03	$72	
Total labor			$210			$210			$392	$812
Provisions and supplies for labor			$70			$70			$95	$235
Hay										
Grain							4 tons	20	$80	
Cottonseed cake										
Salt							2 tons	9	$18	
Total livestock feed									$98	$98
Land leases	6 secs.	$35	$210							
Grazing fees										
Water rental										
Total leases			$210							$210
Auto and truck repair			$25			$25			$25	
Gas, oil, and grease			$30			$30			$30	
License	2	$7.50	$15							
Total auto and truck expense			$70			$55			$55	$180
Real estate taxes										
Personal property taxes										
Total taxes										

FORM A *continued*

Item	January			February			March			Three-Month
	Number or Quantity	Rate	Amount	Number or Quantity	Rate	Amount	Number or Quantity	Rate	Amount	Total Amount
Seed and threshing	…	…	…	…	…	…	…	…	…	…
Twine and sacks	…	…	…	…	…	…	…	…	…	…
Branding fluid	…	…	…	…	…	…	…	…	…	…
Poison, dip, and drugs	…	…	…	…	…	…	…	…	…	…
Insurance	All bldgs.	…	$46	…	…	…	…	…	…	…
Association dues	…	…	$15	…	…	…	…	…	…	…
Building supplies	…	…	…	…	…	…	…	…	…	…
Fencing supplies	…	…	…	…	…	…	300 posts	$.15	$45	…
Machinery repair	…	…	…	…	…	…	…	…	…	…
Blacksmithing	…	…	…	…	…	…	…	…	$35	…
Freight and hauling	…	…	…	…	…	…	…	…	…	…
Camp equipment	…	…	…	…	…	…	…	…	…	…
Fuel	20 tons	$8	$160	…	…	…	…	…	…	…
Feed grinding	…	…	…	…	…	…	4 tons	$3	$12	…
Total general ranch expense	…	…	*$221*	…	…	…	…	…	*$92*	*$313*
Buck purchase	…	…	…	…	…	…	…	…	…	…
Other livestock purchases	…	…	…	…	…	…	…	…	…	…
New equipment	…	…	…	…	…	…	…	…	$215	…
Interest	…	…	…	…	…	…	…	…	…	…
Mortgage payments	…	…	…	…	…	…	…	…	…	…
Note payments	…	…	…	…	…	…	…	…	$75	…
Account payments	…	…	…	…	…	…	…	…	…	…
Total loan payments	…	…	…	…	…	…	…	…	*$290*	*$290*
Operator	…	…	…	…	…	…	…	…	…	…
Personal	…	…	$180	…	…	$180	…	…	$180	$540
Total cash expenditures	…	…	*$961*	…	…	*$515*	…	…	*$1202*	*$2678*

172

Form B. Estimates of Cash Receipts for a Sheep Ranch

Item	January			February			March			Three-Month Total Amount
	Number or Quantity	Price	Amount	Number or Quantity	Price	Amount	Number or Quantity	Price	Amount	
Wool	...	...	...	...	...	...	...	...	...	...
Lambs	820 head on feeding net contract	$4.90	$4018	...	...	...	...	...	...	$4018
Ewes	...	...	...	...	...	...	...	...	...	...
Bucks	...	...	...	...	...	...	...	...	...	...
Pelts	...	...	...	...	...	...	...	...	...	...
Cattle	...	...	...	...	...	...	...	...	...	...
Other livestock	...	...	...	...	...	...	...	...	...	...
Poultry	...	...	...	...	...	...	...	...	...	...
Dairy products	...	...	...	...	...	...	...	...	...	...
Hay	...	...	...	...	...	...	...	...	...	...
Grain	...	...	...	...	...	...	600 bu.	$.75	$450	$450
Pasture	...	...	...	...	...	...	...	...	...	...
Notes and accounts receivable	...	...	...	Note of 2-1-42	...	$350	...	...	...	$350
Borrowings	...	...	...	...	...	...	...	...	...	...
Total cash receipts	...	...	*$4018*	...	...	*$350*	...	...	*$450*	*$4818*

173

FORM C. BALANCE SHEET ACCOUNTS FOR THE STOCK RANCH

Assets		Liabilities and Proprietorship	
1. Cash balance, first of year. $	47.15	1. Chattel notes .... $	3,550.00
2. Bank balance, first of year	685.30	2. Unsecured notes	...
3. Bonds, notes, and shares		3. Accounts payable	76.70
owned	...	4. Leases due	...
4. Accounts receivable	37.50	5. Wages due	...
5. Prepaid wages	110.00	6. Taxes due	377.18
6. Prepaid leases	165.00	7. Advances received on sales	
7. Sales contracts receivable.	718.20	contracts	250.00
8. Sheep inventory, start of		8. *Total current liabilities*	
year	9,630.00	(lines 1 to 7)	*4,253.88*
9. Cattle inventory, start of			
year	750.00	9. Real estate mortgages...	3,255.00
10. Other livestock inventory,		10. Real estate contracts ...	1,250.00
start of year	675.00	11. Equipment notes and	
11. Wool	...	contracts	675.00
12. Hay	1,530.00	12. *Total fixed liabilities*	
13. Grain	365.00	(lines 9 to 11)	*5,180.00*
14. *Total current assets*		13. Owner's investment (assets	
(lines 1 to 13)	*14,713.15*	minus liabilities)	22,969.62
15. Deeded land	8,350.00		
16. Land under contract	2,675.00		
17. Building and improve-			
ments	3,775.00		
18. Machinery and equipment	2,890.35		
19. *Total fixed assets*			
(lines 15 to 18)	*17,690.35*	14. *Total liabilities and propri-*	
20. *Total assets*		*etorship* (lines 8, 12, and	
(lines 14 and 19) *$32,403.50*		13) *$32,403.50*	

174

	1 Sheep	2 Wool	3 Cattle	4 Hay	5 Grain	6 Total
Income						
1. Closing inventory	$6,750.00		$1,875.00			$ 8,625.00
2. Sales	2,773.45	$2,070.18	677.93			5,521.56
3. *Total* (1 and 2)	*9,523.45*	*2,070.18*	*2,552.93*			*14,146.56*
4. Purchases	755.60		275.00			1,030.60
5. Opening inventory	6,895.00		1,655.00			8,550.00
6. *Total* (4 and 5)	*7,650.60*		*1,930.00*			*9,580.60*
7. *Gross income* (line 3 minus line 6)	*$1,872.85*	*$2,070.18*	*$622.93*			*$4,565.96*

Expenses

8. Labor $877.35
9. Provisions and supplies 216.53
10. Shearing 121.15
11. Feed and salt 89.60
12. Leases and grazing fees 133.55
13. Auto and truck expense 186.96
14. Taxes 205.34
15. General ranch expense 266.19
16. Depreciation on equipment 119.88

17. *Total operating expense* (lines 8 to 16) *$2,216.55*

18. *Net income* (line 7 minus line 17) *$2,349.41*

FORM E. WORK SHEET FOR CLOSING THE ACCOUNTS AND PREPARING THE FINANCIAL STATEMENTS

Ledger Accounts	Balances at Close of Year		Adjusting and Closing Entries		Profit and Loss		Balance Sheet	
	Dr	Cr	Dr	Cr	Dr	Cr	Dr	Cr
Bank account	$1,675.15						$1,675.15	
Prepaid wages			(1) 85.50				85.50	
Cattle inventory*	12,115.00		(2)12,365.00	(3)12,365.00†	12,115.00	(2)12,365.00	(3)12,365.00	
Hay inventory*	1,600.00		(4)1,750.00	(5)1,750.00†	1,600.00	(4)1,750.00	(5)1,750.00†	
Real estate inventory*	8,250.00						3,250.00	
Equipment inventory	1,875.00			(6) 105.15			1,769.85	
Notes payable		3,550.00						3,550.00
Owner's investment		20,965.15	(7) 2,271.99					22,270.00§
Owner's personal acc't	2,271.99			(7) 2,271.99				
Cattle sales		5,753.75				5,753.75		
Cattle purchases	335.25				335.25			
Labor	1,060.30			(1) 85.50	975.30			
Provision and supplies	275.95				275.95			
Feed and salt	165.20				165.20			
Leases and grazing fees	88.35		(8) 75.50		163.85			
Auto and truck expense	105.10				105.10			
General ranch expense	320.81				320.81			
Taxes	130.30				130.30			
Depreciation on equipment			(6) 105.15		105.15			
Unpaid leases				(8) 75.50				75.50
Total	$30,268.90	$30,268.90	$16,653.14	$16,653.14	$16,291.91 3,576.84‡ $19,868.75	$19,868.74 $19,868.75	$25,895.50	$25,895.50

*At start of year.   †Closing inventories.   ‡Net profit, credited to owner's investment.
§The increase in the owner's investment is the amount of the net income ($3,576.84) minus the amount of the owner's personal withdrawal ($2,271.99).

176

FORM F. ACCOUNTING FOR CASH RECEIPTS AND CASH PAYMENTS WITHOUT THE USE OF LEDGER ACCOUNTS

CASH PAYMENTS

RECORD OF PAYMENT					ACCOUNT TO WHICH PAYMENT IS CHARGED											
1	2	3	4	5	6	7	8	9	10	11	12	13	14	15	16	17
Date	To Whom Paid	What Payment Is for	Check Number	Amount of Payment	Labor	Provisions	Feed and Salt	Leases	Taxes and Interest	Repairs	Gas, Oil, and Grease	General Ranch Expense	Equipment Purchases	Livestock Purchases	Payment on Notes and Accounts	Personal
Jan.7	A.H.Zimmerman	3 ton cotton-seed cake	176	$77.50	$	$	$77.50	$	$	$	$	$	$	$	$	$

CASH RECEIPTS

RECORD OF RECEIPT				ACCOUNT TO WHICH RECEIPT IS CREDITED								
1	2	3	4	5	6	7	8	9	10	11	12	13
Date	From Whom Received	Received for	Amount of Receipt	Sheep and Wool Sales	Cattle Sales	Other Live-stock Sales	Hay and Grain Sales	Miscellaneous Sales	Sales Contracts	Notes and Accounts Receivable	Borrowings on Notes Payable	General
Oct.16	E.L.Dickson	365 lambs 7¢¢	$1,788.50	$1,788.50	$	$	$	$	$	$	$	$

177

Form G. Livestock Inventory and Production Record

SHEEP

Winter Bands

	Band No. 1		Band No. 2		Band No. 3	
	Numbers	Value	Numbers	Value	Numbers	Value
Counts at Start of Breeding						
Yearlings						
Twos						
Threes						
Fours						
Fives						
Six and over						
Bucks						
Ewe lambs						
Total						
Breeding dates						
Dates to lamb						
March 1 ewe count						
Winter loss, ewes						
Ewe count at lambing						
Number of dry ewes						
Number of lambs docked						
Shearing count, ewes						
Fleece weights						

Summer Bands

	Band No. 1	Band No. 2	Band No. 3
Counts onto Summer Range			
Ewes			
Lambs			
Counts at Marketing			
Ewes			
Lambs			
Summer Loss			
Ewes			
Lambs			
Number lambs sold			
Lamb weights			
Number lambs kept			
Number ewes culled			

CATTLE

Counts at Start of Winter Season	Number	Value
Cows		
Heifers (Long yearling)		
Heifer calves		
Steer calves		
Long-yearling steers		
Two-year-old steers		
Three-year-old steers		
Bulls		
Dairy animals		
Total cattle		
Work horses		
Saddle horses		

Counts at Start of Breeding Season	Number
Cows	
Two-year-old heifers	
Yearling heifers	
Yearling steers	
Two-year-old steers	
Three-year-old steers	
Bulls	
Females per bull	
Breeding dates	
Production and Sales	
Number calves branded	
Number calves raised	
% Calf crop, cows	
% Calf crop, heifers	

Death Losses for Year	Number	Weight
Cows		
Heifers		
Steers		
Calves		
Bulls		
Sales		
Calves		
Long-yearling steers		
Long-yearling heifers		
Two-year-old steers		
Three-year-old steers		
Cows		
Bulls		

FORM H. FEED-CROP INVENTORY AND FEED USE RECORD

	Alfalfa Hay Am't	Value	Native Hay Am't	Value	Grain Hay Am't	Value	Wheat Am't	Value	Oats Am't	Value	Barley Am't	Value	Cottonseed Cake Am't	Value
Amounts on hand at start of feeding														
Bought during feeding season														
Sold during feeding season														
Fed during feeding season														
Carried over at end of feeding season														
Raised during crop season														
Sold during summer and fall														
Bought during summer and fall														

	Calves	Cows	Heifers	Steers	Sheep, Band	Sheep, Band	Horses, Band
Dates started winter feeding							
Rates of feeding:							
Alfalfa hay							
Native hay							
Grain hay							
Grain							
Cottonseed cake							
Dates stopped feeding							
Days winter grazing							
Total days on feed							
Total amounts of feed fed:							
Alfalfa hay							
Native hay							
Grain hay							
Grain							
Cottonseed cake							

FORM I. BUILDING AND EQUIPMENT INVENTORY

	Buildings							
	Size	Original Cost or Value	Esti-mated Life	Age	Value First of Yr.	Repairs This Yr.	Deprecia-tion for Yr.	Value End of Year
Sheds								
Shelter								
Barns								
Fences								
Corrals								
Water development								
Dwellings								

	Equipment							
	Size	Original Cost or Value	Esti-mated Life	Age	Value First of Yr.	Repairs This Yr.	Deprecia-tion for Yr.	Value End of Year
Haying equipment								
Farm machinery								
Trucks								
Tractors								
Autos								
Shearing								
Camp								
Saddle and harness								

181

Form J. Inventory of Notes and Accounts Receivable

	Amount First of Yr.	Date Due	Interest Rate	Rec'd during Year	Amount End of Year
Land mortgage notes					
Land sale contracts					
Chattel notes					
Unsecured notes					
Open accounts					
Prepaid leases					
Prepaid taxes					
Prepaid wages					
Advances made on purchase contracts					

Form K. Inventory of Notes and Accounts Payable

	Amount First of Yr.	Date Payable	Interest Rate	Paid during Year	Amount End of Year
Land mortgage notes					
Land purchase contracts					
Chattel notes					
Unsecured notes					
Supply accounts					
Unpaid leases					
Unpaid taxes					
Unpaid wages					
Advances received on sales contracts					

FORM L. LAND INVENTORY AND RANGE USE RECORD
(Space at the left of this form is for a map of the ranch)

Land Inventory

	Acres	Value
Irrigated Cropland		
Dry Cropland		
Native Meadowland		
Fenced Pasture		
Spring-Fall Range		
Summer Range		
Winter Range		
Total Value (Deeded Land)		

Spring-Fall Range Use

Kind of Livestock	Number of Head	Date On	Date Off	Days of Use	Estimated Grazing Capacity

Summer Range Use

Kind of Livestock	Number of Head	Date On	Date Off	Days of Use	Estimated Grazing Capacity

Winter Range Use

Kind of Livestock	Number of Head	Date On	Date Off	Days of Use	Estimated Grazing Capacity

183

Month	Name of Employee	Days of Month 1-31		Total Days Work	Supplies Drawn	Amount Paid
June	J.D.James	X X X X X X X X X X X X X X X X X X	X X X X X X X X X X X	25		$75

End of Month Estimate of Time Allocation

Month	Name of Employee	Sheep	Cattle	Hay		Ranch Development		Kind of Equipment Used
				Man days	Equipment days	Man days	Equipment days	
June	J.D.James			13		12		Buck rake Fresno

184

Cattle

Breeding cows (196)	$11,760
Heifers (10)	500
Yearlings (137)	4,795
Total cattle	$17,055
Approximate average per head	$50

Work Stock

Saddle horses (6)	$360
Work horses (12)	600
Total work stock	$960
Approximate average per head of cattle	$3

Land

Grazing land (8,200 acres)	$16,400
Native meadow (125 acres)	2,500
Dry cropland (160 acres)	1,600
Total land	$20,500
Approximate average per head	$60

Feed

Native hay (150 tons)	$1,050
Grain hay (100 tons)	500
Wheat (1000 bushels)	700
Total feed	$2,250
Approximate average per head	$6.50

Equipment

Mowers (2)	$180
Rakes (2)	90
Stacker	175
Hay racks (2)	50
Header (1)	125
Tractor	600
Tractor plow	115
Discs	75
Harrow	60
Drill	185
Grain binder	195
Wagons (3)	150
Feed grinder	160
Gas engine	165
Saddles (3)	120
Harness (6 sets)	360
Pick-up truck	500
Auto	600
Small tools	275
Total equipment	$4,180
Approximate average per head	$12

Buildings

Barn (1)	$600
Granary (1)	250
Repair shop (1)	400
Machine shed (1)	350
Dipping vat (1)	250
Corrals	250
Bunk house	400
Garage	150
Total buildings	$2,650
Approximate average per head	$8

Total investment $47,595.00

Total investment per head of cattle .. $139.50

CHAPTER VII

Planning the Management Program of the Ranch

A successful ranch operator plans beyond the current year. He has a program which encompasses his future management hopes and goals and serves as a background for the annual production planning. Let's have a look at the most important of the management features which should enter into a well-rounded future program for the ranch.

THE SIZE OF THE RANCH

Most important among the factors determining the best size for a ranch are (1) the number of livestock needed for good technical efficiency in livestock management, (2) the resource geography as it affects the size and balance of ranch operations, and (3) the management capacities and limitations of the ranch operator.

Size for Livestock Management. If a cattle ranch is to attain high technical operating efficiency, it needs 300 to 500 head of cattle. This size usually permits full exercise of the opportunity for classification of the stock for management, for use of grazing lands with special values or adaptations (such as a steer range or a cow-and-calf range), for economy in the use of feed crops, and for the use of specialized labor. Beyond 500 head there is not, as a rule, much to be gained in the way of technical operating efficiency by increasing the number of cattle operated.

For the sheep ranch that uses winter range throughout the winter months, a size of 2000 to 2500 ewes (including the yearlings kept for replacement) means good technical operating efficiency. This number will be operated as two summer bands and one winter band. The sheep ranch that has to practice winter feeding requires from 3000 to 3500 ewes to attain good technical operating efficiency. These numbers will be operated as three summer bands and two winter bands. However, the optimum in technical operating efficiency for the sheep ranch is realized considerably beyond these numbers. This size is about 6000 to 7000 ewes where the rangelands and croplands can be adjusted for a unit of this size and the managerial capacity of the operator is sufficient to handle it. The sheep ranch of such size can classify the ewe bands by age from yearlings to six-year-olds and can operate a separate band of yearlings for replacement. Such classification permits an adaptation of the various age bands to the different range conditions which may prevail, and to the different types of winter feed which may be available.

Resource Limitations on Size. The local geography of the range and cropland resource may be a limiting factor in the economy of the size of the ranch and, in some situations, largely determines the size of the operation. For example, the size of the foothill ranches and the mountain valley ranches of the Rocky Mountain region is often predetermined by the availability of natural land units of seasonal rangelands, irrigation water, and croplands. A natural unit may not lend itself to subdivision into smaller units, or to combination with other units. In planning for any expansion or decrease in the size of the ranch as a land unit, it is, consequently, necessary to analyze the resource geography in terms of ranch management possibilities. Another illustration of the influence of resource geography on the economy of the size of the ranch is the highly seasonal sheep ranches of the intermountain region. Seasonal migration of considerable distance is necessary to use these ranges, and the size of the operation may be influenced or even predetermined by the balanced unit of seasonal rangelands that is attainable within the limits of the migration possibilities.

As a generalization, it may be said that the efficient size of the ranch is the result of the best adaptation and compromise between technical efficiency in livestock management, the local geography of the land resource ("the way the country is put together"), and the management capacities and interests of the operator. There are instances where, because of the superior business management ability of the operator, the size of the ranch has been developed and maintained beyond best technical operating efficiency.

Present Size of Stock Ranches. A majority of the western stock ranches are, however, below the size of optimum technical operating efficiency. Especially is this true for the cattle ranches. Primarily, the cause of this is the relationship between the supply of resources and the number of people who wish to make ranch ownership and operation their life endeavor. Western ranch economy is probably not far different in this respect from the general national picture of agricultural economy and farm management, except that the small stock ranch enterprise is less adaptable for family living than is crop agriculture.

It is, consequently, necessary for many ranches to operate below the level of best technical efficiency in the management of livestock and in the use of lands and labor. For these ranches, the major emphasis in planning the size of the operation should be on attaining an income that will permit a reasonable standard of living.

A small ranch with a capacity for 150 head of cattle may be expected to yield a net income for family use averaging about $1000 annually if there is no debt service. The small sheep ranch that operates one band of 800 ewes can expect an average annual income for family use of about $1000 if there is no debt service.* Sometimes these small ranches can develop a supplemental enterprise of crop production and livestock fattening, poultry, or dairy production. These enterprises depend on the adaptability of the resources and the availability of family labor. However, where supplemental enterprises are planned, a careful consideration of the availability and suitability of the resources is needed. Usually the approach should be to test out any such enterprise in a small way at the start and "grow into" it.

LAND OWNERSHIP AND TENURE

Tenure Methods. As a rule western stock ranches control their lands through some combination of deeded land, land purchase contracts, leasing of state, county, and privately owned lands, and public land grazing permits. Consequently, a majority of ranches have some problem of management planning in choosing the means for stable control of lands and irrigation water rights. The ranches of western Texas where land ownership is generally well

* This is in terms of 1935–39 prices and ranch operating costs.

consolidated in the ranch units are an exception to this. There, the lands did not pass into private holdings through the usual homestead entry, and there is but little federal public land.

In the management planning for land ownership and tenure a base map of the ranch showing the land ownership and tenure status by ownership tracts is desirable for the medium-size ranch and a necessity for the large ranch. This type of map affords a visual aid in planning for land control and stability of the land unit of the ranch. Plans and decisions must be made on relative priority for the purchase of leased lands in the ranch unit, and on what lands should be most secure in tenure for blocking up and rounding out the land unit for management. This map is also useful for planning the strategy of land control to lower the "nuisance value" of some lands to competing operators, who might undertake to control leased lands that are a valuable part of the ranch unit.

Costs of Owning and Leasing. Competition for rangelands is forcing the western stock ranch more and more into rangeland ownership. This may eventually cause a considerable change in the present situation of desert and semidesert ranches that own only the "key lands" in the unit and lease a substantial part of the rangelands from private holders, from the state, county, and federal government, and from large institutional owners such as railroads and land credit agencies. There are a good many such ranches that own but little of the land they use.

However, the present relationship between the costs of ownership and the cost of leasing rangeland is such that a combination of owning and leasing is desirable. This is especially true where it is possible to secure term leases on privately held lands and term grazing permits on administratively managed public lands. A ranch with all deeded land and carrying a heavy debt service and high taxes is likely to have less secure tenure than the ranch which depends in part upon leases, particularly term leases that can be protected against excessive competition.

Deeded Lands. For the ranches that require the use of croplands in the year's operations, the cropland is nearly always deeded land. This is necessary for stability in the control of rangeland leases and for holding public land grazing permits. In the regions and localities where winter feeding is not necessary and the ranges are seasonal, the deeded land of the ranch unit is, as a rule, the seasonal range that is most limited in supply, because such rangeland is the most valuable property base for leasing other lands and for obtaining grazing permits on the seasonal ranges in public ownership. This situation is illustrated by the extensive ownership of the spring-fall ranges by the sheep ranches of the intermountain region, and the recognition of such ownership as satisfying the property needs or "deeded land base" for grazing permits on the national forest summer ranges and on the grazing district winter ranges. In the more arid regions and localities where livestock water is the key resource, the lands with water development are the deeded lands and form the nucleus of the ranchland unit for leases and public land permits.

Irrigation Water Ownership. For the ranch that uses irrigation water in the production of feed crops or pasture, the ownership of adequate water rights is essential. Throughout the western states the water rights to stream flow are based on appropriation and use. In several states the rights are definitely established by court decree. In others court decrees have established rights on streams only when litigation arose between parties claiming water. In such states the doctrine of appropriation and use is recognized, even though there never has been a formal filing by the user. In all western states earlier appropriations take precedence over later appropriations in conformity with the principle "First in time, first in right." Thus in the localities where the irrigation water requirements exceed the supply, the earlier rights have priority as the natural stream flow diminishes through the summer. Adjudicated water rights can often be transferred by sale. Consequently a ranch with inadequate water supply sometimes may acquire supplemental rights through purchase if any water can be found for sale in the locality.

In planning for such acquisition, the stream flow records and the relationship of the different adjudications of the rights to the flow should be examined in order to determine whether the water right that might be purchased is of such priority that would meet the usual requirements of the ranch. The amount of water that would be available from the right during a year of low stream flow, particularly for the latter part of the season, should be determined.

Ranch irrigation water rights usually provide for direct intake from small streams. (Montana Engineer Photo)

There are several different types of organization engaged in the development and delivery of irrigation water. Probably the majority of western ranches obtain their water by directly diverting it from a stream flowing through or near the property. In these instances the cost of water is only a nominal amount of labor each year for repairing any diversion structure and maintaining the ranch ditch. It is not uncommon for two or more ranches to divert at the same point and to maintain jointly any diversion structure or canal on a partnership basis. Where large diversion dams, storage reservoirs, and/or expensive canals are involved, an irrigation company or an irrigation district is usually formed to administer the enterprise.

Under an irrigation company each user ordinarily owns shares of stock in proportion to the acreage he irrigates. Annual assessments are made against these shares to defray maintenance and operating costs and for the retirement of any debt. An irrigation district is a political subdivision of the state. Its annual charges are assessed against all land within the district and often are collected by the county treasurer along with general property taxes. It is important to note that all lands in an irrigation district are jointly liable for repayment of any debt owed by the district.

Where the ranch does secure irrigation water through an organization that has constructed storage facilities and/or delivery canals, the charge for the water should be analyzed in terms of the cost per ton of hay for the normal yield. Generally this cost should not go beyond a maximum of $1 per ton of normal hay yield for the stock ranch. There are instances where the charge for water from an irrigation district or company will run as high as $6 or $8 an acre for the operation and maintenance cost, and in addition there may be a construction indebtedness that approximates the full normal value of the irrigated land. These extreme situations of water cost necessarily have an adverse effect upon the sale value of the ranch property that has contracted for such high-cost water. The water facility debt and the annual operation and maintenance charge, as well as the adequacy of the water supply and the priority of the water right, should, therefore, receive careful attention by one purchasing a ranch or a cropland addition to a ranch.

RANCH AND RANGE IMPROVEMENTS

A program of adequate and efficient improvements for the management of the ranch concerns principally: (1) the plan of fencing for the rangelands, meadows, and pastures; (2) range water development; (3) livestock management facilities, such as range camps, corrals, pens, dipping vats, chutes, barns, and sheds; (4) roads and trails; and (5) the plan and layout of the ranchstead for the location and arrangement of buildings.

The Fencing Plan. The plan of rangeland fencing to serve the dual need of range and livestock management should be based upon an analysis of management needs and opportunities. Examples of such possible needs and opportunities are seasonal rotation grazing of different parts of the rangelands, fencing for distribution of livestock on the range and for efficient use of water facilities in range and livestock management, and the fencing of range pastures for such management purposes as winter feeding, calving, lambing, and beef finishing.

Planning for Water Development. Well-planned water development reduces to the practical minimum the distance the livestock must travel to water and facilitates the distribution of livestock use over the range. The kind and cost of water facilities will vary with the opportunities for water development. Enclosing a spring and piping the water into a trough is the lowest cost type. Fenced earthen reservoirs with the water piped into a tank can be an efficient and low-cost development. Power-operated wells with storage tanks from which the water is piped to troughs have the highest cost.

Excepting the southern plains and southwestern regions and some parts of the intermountain region, it is seldom necessary or desirable to invest in water facilities more than $1 per animal month of rangeland capacity (for cattle). The cost of the minimum need for water development may be twice or even three times this amount in the regions mentioned.

Livestock Management Facilities. Planning the needed facilities for livestock management should center around the most efficient place for their location and economy in the use of labor time and the movement of livestock. For example, a branding corral for cattle or a dipping vat for sheep should be placed, if possible, on the part of the range where the livestock is located at the time when these practices should be performed. Range camps should be located where they are most accessible to the rangeland management units which the camps serve. Pens for sorting and holding livestock usually should be located at or close to the ranchstead.

Ranch and Range Roads and Trails. Roads and trails should be planned for the easiest access by car, truck, or horse to different

Location and plan of the ranchstead can be an important feature in management. (Oliver Brothers Ranch Photo)

parts of the ranch from the ranchstead, in order to observe the livestock, supply the camps, service the water facilities, and repair the fences.

Location and Arrangement of Ranch Buildings. Ranchstead locations and plans, once established, are not easily changed. Any one of several factors may predominate in determining the location — the availability of good water, accessibility to highway transportation, the desirability of a location close to the lands where the feed crops are produced and the livestock fed through the winter. For the larger ranches, which employ considerable labor for the production of feed and the handling of livestock, the location and plan of the ranchstead should facilitate accessibility to and supervision of the operations that have high labor requirements. Where the ranchstead is located on a stream drainage the ranchstead should be up-drainage from the meadowlands, with the barns, sheds, and the like, down-drainage from the site of ranch dwellings and the supply and storage buildings. Ranchsteads should be on a southern exposure in northern climates.

IMPROVEMENT OF RANGELAND AND CROPLAND

CAPACITY AND BALANCE

The management plan of the ranch for livestock numbers and production undertakes to make a comparison of the present capacity of the ranch with what is considered an economically attainable objective. For the grazing lands this comparison is based upon the use history of the lands and the condition and trend in the condition of the range, to show the present range acreage requirements per head and the potential improvement in capacity through management. This analysis of the present and potential capacity of the ranch should not be *only* in terms of livestock numbers, but in terms of numbers *plus the tonnage and quality of market turnoff resulting from rangeland and cropland improvement.*

Rangeland Capacity. Plans to meet management needs and to realize possibilities for improving the capacity of the rangelands should be made in terms of concrete objectives to meet a problem. Such improvement may be accomplished by rotation management of a range, reseeding some parts of it, or other applicable methods.

A New Mexico cattle ranch, for example, that now has an average grazing land acreage requirement of four acres per animal month has undertaken a plan of deferred and rotation grazing, spreading water runoff on some parts of the range, and reducing the present animal months of grazing use by about 20 percent. Within a period of five years the plan should, on the basis of experimental results, reduce the average acreage requirement to three acres per animal month and substantially increase the total quantity and the quality of the market turnoff. Financial calculations based on estimates of production results indicate that the present value of the increased future revenue from this improvement will be about double the cost of the improvement.

Planning for improvement in the capacity of grazing lands, or for the addition of grazing lands to the ranch unit, should always be based upon an analysis of the ranch's need for a better balance of rangeland capacities through the season; and also, in some cases, for a better balance between the rangeland capacity and the feed-crop capacity of the ranch.

Feed-Crop Capacity. In making plans for the improvement of the feed-crop capacity of the ranch, there are two factors to consider in the capacity estimates: (1) the usual length of the winter feeding season and the feeding rates to determine the feed requirements per head; and (2) the acreage, yield, and production of the feed crop (the capacity of meadow pasturage and crop feed aftermath should not be overlooked in making capacity estimates for the feed-crop lands). A comparison of these two factors will show the probable result, in terms of capacity for maintaining livestock through the winter feeding season, of any possible plan for improving feed-crop lands or for introducing more productive types of feed crops.

Rangeland and Feed-Crop Balance. The estimates of the potential feed-crop capacity can then be compared for probable balance with the plans of future improvement in the capacity of the rangelands. Rangeland capacity is estimated by multiplying the probable average rangeland acreage requirement per animal month by the usual length of the grazing season (to get the grazing season acreage requirement per animal) and dividing this result into the available acreage of rangelands. This gives the number of head which can be operated for the grazing season. If the ranch uses two or more seasonal ranges, this estimate will be made for each seasonal range.

A ranch operator is expressing his idea of the capacity balance of his present operation when he says, "This ranch will graze two hundred head of cattle and winter two hundred fifty head." But the results of his experience may be only a partial basis for forecasting the probable results of a change in the management program of the ranch. Past standards of range and feed-crop use may not be applicable to management opportunities made possible by new types of feed crops and new range improvement methods.

LIVESTOCK MANAGEMENT PRACTICES

Plans for livestock management practices as a part of the management program of the ranch will usually emphasize adaptation of breed and type of animal to market outlook and demand, factors influencing calf and lamb crops and animal death losses, and practices affecting market weights.

Breed and Type of Animal. Livestock breed improvement and the development of the type of animal best suited to the resources of the ranch and to the market outlook center upon the selection of sires and replacements for the breeding herd. There are few ranches that have approached the limit of their possibilities for breed improvement to attain higher growth efficiency and better quality and greater uniformity in market grades.

The choice of the cattle ranch in breeding and selecting for the type of beef animal is, principally, a choice between the beef breeds and types that will produce good young feeder animals, which will make efficient gains in the feedlot, and the breeds and types that are somewhat slower in maturing but able to make more efficient use of rough, dry, or low-capacity rangelands. As a general rule, the Angus breed and the blockier and earlier maturing types of the Hereford breed are best adapted to high-capacity range that approximates pasture conditions. The large and "rangy" types of the Hereford breed generally make better use of foothill and rough upland range and are better adapted to "roughing through" the winter feeding period. Consequently, the cattle ranch operator, in breeding for a type of Hereford, Angus, Shorthorn, or other breed or cross, must analyze the suitability of his range and crop feeds for that type, and the prospective market demand and prices for the various types and ages of beef animal.

The breeding program of a sheep ranch must be based upon the relative adaptability of the range and crop feed combination of the ranch for the mutton breeds and the fine-wool breeds, and upon the market outlook for the different classes of lambs and grades of wool. The breeding program of the plains sheep ranches usually concentrates on the fine-wool breeds, such as the Rambouillet, and aims to secure stability and uniformity for wool and feeder-lamb production.

Sheep ranches of the Rocky Mountain and intermountain regions having upland summer range should aim at a breeding program that gives a good combination of slaughter-lamb production and wool production. It appears probable that these ranches will do less crossbreeding in the future than they have in the past. Too frequently the result of unplanned crossbreeding of fine-wool and mutton breeds has been a lack of uniformity and stability in the type of lambs and wool produced. The alternative for these ranches is to use mutton-breed bucks with a large type of Rambouillet ewe and to buy the replacements, or to use the Columbia breed or other crossbred whitefaced breeds.

Production Practices. A sheep ranch with an annual average lamb crop (defined as the ratio of lambs raised to the number of ewes at breeding time) below 85 percent, and a ewe death loss above 6 percent, should make a careful analysis of its livestock management practices that influence crops and losses. This same observation applies to the cattle ranch where the calf crop (the ratio of calves weaned to the number of females of breeding age at breeding time) averages below 75 percent, and the annual death loss averages above 2 percent.

Livestock management practices having the greatest influence on lamb crops are conditioning or "flushing" the ewes before breeding, supplying an adequate number of vigorous sires, culling ewes that are poor breeders, providing adequate labor and care to save the largest possible number of lambs at lambing time, and good herding to avoid loss of lambs on the summer range through strays, predators, and the like. The livestock management practices which should receive principal attention in planning to increase calf crops are supplying an adequate number of sires and effectively using them by keeping them distributed on the breeding range, culling sires and females that are poor breeders, providing observation and care for the breeding herd during calving, and preventing disease, such as contagious abortion and blackleg.

Starvation death losses are now of less concern in the management planning of the stock ranch than has been the situation in the past, when speculative and hazardous types of operations resulted in heavy losses during severe winters and prolonged summer droughts. A good program of range management and adequate winter feed supplies are the best insurance against climatic risk. Disease, accident, theft and straying, and poison plant losses are the other principal causes of death loss that warrant attention in analyzing and planning the livestock management program of the ranch.

Market Weights. Livestock management practices that deserve the most attention in analyzing the opportunities of the ranch to increase the weight of animals marketed are making plans for the adequate use of ranges, pasturage, and feed crops; breeding for efficiency in gain and growth rates; and timeliness in marketing to avoid the low-gain rates and loss of "bloom" that may occur during the latter part of the season.

Since most of the year's gain for the market livestock is made on the range, livestock management to make the most effective use of range forage is most significant in securing good market weights. The method of herding the ewe-and-lamb bands on the summer range, such as open herding rather than close herding, can make significant differences in the market weights of lambs. The extent to which cattle are kept well distributed to secure even use of a range through riding, salting, and well-planned water development can be an important factor in gains and market weights. Where the feed resources of the ranch make it possible, a part of the summer range may be profitably managed as a finishing range for the market livestock. It is good economy for some ranches to use the hay meadow regrowth during the late summer and early fall to put additional gain and finish on the market animals.

The matter of breeding and selection to attain efficiency in growth rates and market weights should not be regarded as purely a livestock management problem. It is a problem in adapting breed and selection of type to the range and feed-crop management opportunities of the ranch. The ranch that has made such an adaptation will have both a favorable *average* weight of market animals and a low variation of individual animal weights from the average weight.

In considering management plans for selling market animals to make the best use of weight gains, it can be stated as a general rule that when the rate of gain for cattle falls to a late-season average of a half pound per head per day, the economy of the gain is questionable. There are several ranch management aspects of this to be considered. It may be good economy to sell the market animals before gains fall to this point if the range feed is going to be needed to maintain the breeding herd at the efficient level.

On the other hand, it may be desirable to retain the market animals after their rate of gain has dropped below this point if crop aftermath and meadow pasturage will become available later in the season for making added gains and finish during the fall months. The timing of the marketing is of particular importance to the sheep ranch that markets slaughter lambs from the range. If a high percentage of the lambs are to sell as slaughter animals, they must be marketed before they lose the young-animal fat resulting from the combined diet of milk and succulent range feed.

PLANNING FOR POSSIBLE CHANGES IN KIND OF LIVESTOCK

Excepting the situations where the type of range and the feed crops clearly limit the ranch to one or the other kind of range livestock, most ranch operators have at one time or another considered the adaptability of their ranch to cattle or to sheep, or to a combination of both. As a ranch management planning method of considering such possibilities, it is a desirable procedure to make some systematic budgetary calculations and estimates regarding the alternative production possibilities of the ranch.

Capacity Comparisons. The first of these estimates should compare the rangeland and feed-crop capacities of the ranch for the two kinds of livestock, or for a combination of both kinds of livestock if that appears to be a possibility. Where the rangelands do not have any use history for one or the other kind of livestock, the capacity estimate for the grazing lands will have to rest on the known use history of similar lands. The capacity estimates should stress any possible lack of balance in the range and feed-crop capacities that may result from a change in kind of livestock, and how such a lack of balance might be offset by adding range, changing the seasonal use of range, shifts in feed-crop production, or buying feed.

Income Comparisons. Based on these capacity estimates, calculations can be made about probable marketable production. Market prices can then be applied to these estimates to compare probable gross incomes from the different programs that appear to be feasible. The next step in making the estimates is to calculate the probable physical requirements for labor, equipment, purchased feeds, leased lands, and the like, and to apply cost rates to these requirements to get operating cost estimates. There are many cost elements for which physical requirements cannot be estimated without burdensome detail, and these costs can be estimated directly in monetary amounts. A budgetary form such as that illustrated by Form A can be used in making the cost estimates, and the results can be checked against the average standards shown in Tables 15 and 16. These cost estimates are then subtracted from the income estimates to compare the probable average annual net revenues from the programs that are being analyzed.

It might be said that if the results of such systematic analysis and ranch program planning are clear-cut and significant, the alert and experienced manager will have discovered the same results by trial and error without any such systematic or formal analysis and will be working toward them. But, such systematic and detailed analysis is often valuable, even for the old-time operator who can see no merit in "producing 'em on paper."

MARKETS AND MARKETING METHODS

Recent Changes and Trends. Changes in markets and marketing methods may be an important feature in the future plans for a ranch. One reason for a change in markets is illustrated by the eastward shift, during the past decade, of the transition zone for the eastward and westward market movement of cattle from western stock ranches. Change in marketing methods is illustrated by the recent rise of the country auction market. A probable future change in wool marketing methods is the development of wool shrink and grade analysis at country points, possibly in connection with wool auction markets. It is important that the ranch operator be aware of definite trends that may be developing in market outlets and marketing methods, appraising the significance of such trends in terms of the management program of the ranch. *Long-Run and Short-Run Changes.* In this phase of the management analysis and program planning of the ranch, it should be borne in mind that some changes in markets and market demand are transitory and do not signify any definite trend. This is illustrated by the wartime change in the feeder-cattle markets, when the competition for other uses of feed grains caused a shift in the feeder demand to older animals with a lower degree of finish and widened the slaughter-market outlet for range steers and cows. A change of this type may be very important in making the annual production plan of the ranch, and yet not warrant any adjustments in the long-time features of the ranch, such as breeding and selection for type of animal or the plan of use of rangelands and feed crops. The alert ranch operator keeps currently informed through the various available market news services on price trends and price comparisons for the different central markets and other market outlets available. But he is also thinking in terms of trends in the changing picture of markets and marketing methods, and of the significance of such trends to price advantage in marketing and to management changes in production.

OPERATING COSTS

Adapting Costs to Resources. Planning for operating costs in the management program of the ranch should aim at the economic balance between costs and the productive capacity of the ranch. Ranches that operate on highly productive resources can produce a higher market turnoff *per head of livestock capacity* than those which operate on low-grade lands. Ranches with the more productive resource should have the higher operating costs if they are to maximize their income opportunity. A ranch with low-grade resources may, by adjusting operating methods and costs to the production opportunities of the resources, attain a *net* profit per unit of capacity that is not far below the ranch with the high-grade resources.

This adaptation of costs to the production opportunities of the resources is illustrated by the fact that the mountain valley sheep ranches use approximately twice as much labor as the winter range sheep operations of the intermountain region. There is also a general contrast between the semidesert cattle ranch operations of the intermountain and southwestern regions and the foothill cattle ranches of the Rocky Mountain region. The operating costs of the former type of ranch will average only about 60 percent of the justifiable production costs of the latter type.

In analyzing the operating costs of the ranch and planning for possible changes, it is desirable to compare the present operating cost situation of the ranch with the standards for different types of ranches given in Tables 15 and 16, and to compare the present production performance of the ranch with the standards given in Tables 8 and 11.

Use of Operating Cost Standards. If such a comparison shows any marked differences between the present operating costs of the ranch and the applicable cost standards, the analysis should attempt to determine the reasons for the differences. Such differences may be justifiable because of some unique feature of the operating conditions of the ranch, but they also may be due to a lack of balance between the cost inputs and the production output opportunities of the ranch.

This lack of balance may be caused by underexpenditure as well as overexpenditure. An illustration of this is a mountain valley sheep ranch that has a good opportunity for lamb production but is not realizing the opportunity because of the low input of labor time and feed. The consequence is an uneconomically low operating cost, more comparable to the operating costs of a plains sheep ranch that can produce only 55- to 65-pound feeder lambs than to the economic level of costs for early-lamb production.

RANCH INVESTMENT VALUES AND FINANCE

The Concept of Normal Values. Planning for ranch capital values and land credit should be in terms of a concept of normal market prices and operating costs, recognizing that there will probably be recurrent periods of inflated sale prices and distress sale prices of ranch and farm property.

One method of appraising the ability of a ranch to carry capital values is to estimate the probable normal annual gross income and the operating costs (including a charge for current interest rates on the values of livestock and equipment), to determine the net income available to pay a wage to the operator and an interest return on land values. From this amount the competitive market wage rate for the operator and family labor is deducted and the residual income is capitalized at land loan interest rates to determine the capital value of the ranch property.

This method is not too realistic when applied to the individual ranch. Any errors in the estimates of gross income, operating cost, and resulting net income are multiplied many times in the capitalization process. This method does have merit in the development of ranch valuation standards from the operating data of a large number of ranches of similar size and type, since the use of these group data reduces the chance of error in the results. The per-acre rangeland and feed-crop land value standards given in Tables 1 and 2 were derived by this method. Such standards serve a useful purpose in checking the results of other appraisals for the individual ranch, such as past market sale prices of that ranch and similar ranches.

One method for estimating the probable ability of a ranch to carry land credit is to obtain the operating history of the ranch for these factors: (1) what may reasonably be expected in annual marketable production; (2) the annual gross value of such production with medium prices such as those shown in Tables 7 and 10; (3) the annual operating costs of the ranch with medium cost rates for labor, supplies, equipment, leases, taxes, and the like; (4) the resulting income available for family living and debt service; (5) the annual income requirement for family living; and (6) the residual income available to meet debt service. Land credit agencies have in recent years used such an approach as a means of analyzing the debt service capability of ranch properties which are under consideration for real estate loans.

Superior Management for Increasing Values. The appraisal of the capital value of a ranch property and the analysis of the financial program of a ranch by a land credit agency must of necessity be based upon the present organization and operating methods of the ranch, and upon average quality of management. The program planning of the operator for ranch values and finance can, however, undertake to evaluate the reorganization possibility of the ranch and the possible results of management improvements.

The capital value of a ranch may, for example, be very materially enhanced through a change which makes for better balance in the feed resources or a more efficient size of the operation, and the calculation of ranch capital values by the operator can in some degree be based upon such an improved management program for the ranch.

PLANNING FOR THE RANCH AS A MANAGEMENT UNIT: A SUMMARY

Selecting the Important Management Features. The various features that enter into management planning for the ranch must be considered in relation to each other in the development of an effective management plan. The first step in the development of the plan is to analyze the present management program of the ranch and to select the features of management that should be given primary attention in reshaping the ranch management program. The next step is to consider the possibilities and the alternatives for change in the management features that are selected, and to analyze the probable effect of a change in one management feature upon the other management features of the ranch.

Management Features and Interrelationships. As a procedure toward this end, it is desirable to use a check list of the management features to be considered and from which a selection is made of the principal needs and possibilities of the ranch for reshaping the management program. The following is such a list, with some of the possible interrelationships indicated.

1. Size of the ranch. Does the present size permit good technical efficiency in livestock management, land management, and specialization in the use of labor? Is there any opportunity for expansion or subdivision of the present land setup of the ranch to effect a change in size, and would such a change improve or impair the balance in the seasonal capacities of the ranch?

2. Land ownership and tenure. Is the ranch well balanced between grazing land ownership and leasing, in order to get a good combination of moderate land costs and stability of tenure? Would additional land purchase improve the land control situation, or contribute to plans for ranch and range improvement work?

3. Ranch and range improvements. Are the present improvements adequate for management? What is the present investment

in such improvements and what are the economic limits of the ranch for further investment?

4. Balance of range and cropland capacities. In what season of the year is feed deficiency most likely to occur? Can any present lack of balance be corrected better by a change in management of the present unit or by a change in the land holdings?

5. Livestock management. Does the present plan of breeding and selection give a good adaptation of production to the feed resources of the ranch and to the market outlets? Could a better adaptation be made by a change from sheep to cattle or vice versa or by combining cattle and sheep in the operation?

6. Kind of livestock. Is the ranch inherently suited to cattle or to sheep? What are the relative income possibilities of the ranch for cattle, sheep, or a combination? What change in improvements and equipment would a shift require and what would the cost be?

7. Markets and marketing methods. What trends are occurring in market outlets and marketing methods for the products of the ranch? Do these trends indicate any desirable changes in type of product or in marketing methods?

8. Operating costs. Are the costs balanced with the resource productivity and income opportunity of the ranch? To what extent might any lack of such balance be due to a poor adaption of the kind of livestock and type of production to the range and crop resource?

9. Ranch values and finance. Does the present debt service of the ranch permit adequate income for the operator? Can the income and the capital values of the ranch be enhanced by reshaping the management program and by superior management on the part of the operator?

REPLANNING THE MANAGEMENT PROGRAM OF A RANCH:
A CASE ILLUSTRATION

The ranch which is used here as a case illustration of method in management analysis and planning is located in the Bear Paw Mountains of north central Montana. It is a mountain valley type of sheep ranch, now operating approximately 8000 ewes, including the normal annual requirement of about 1600 yearling ewes. The Bear Paw Mountains are one of the several lower mountain areas located in the northern plains to the east of the Rocky

Mountains. Ranches of the Bear Paw Mountain area are located around and within the mountains. Mountain rangelands and the valley feed-crop lands are under private ownership. These lands constitute the principal deeded lands of the ranches. High plains which slope away from the mountains are the spring-fall range. This land is leased and owned by the ranches.

Our analysis of the present management program of this ranch and our proposals for change in its present program will follow the method given in the preceding summary, "Planning for the Ranch as a Management Unit."

Size of the Ranch. This ranch is somewhat larger than the size needed to attain full technical operating efficiency for this ranch type. The ewes are operated in seven summer bands and four winter bands. Yearling sheep are operated as a separate band. A summer band of ewes with twin lambs is operated. Aged ewes are marketed in the fall as six-year-olds. It appears probable that this ranch could gain some in operating efficiency by reducing the ewe numbers to the point where they could be handled in six summer bands of about 1000 head each and combining these into three winter bands for winter feeding. The present average size of the summer bands is somewhat too large for good management on the mountain range.

Land Ownership and Tenure. This ranch owns 550 acres of irrigated hayland, 250 acres of dry cropland, and 17,000 acres of mountain summer range. The spring-fall range of some 32,000 acres is all leased from private and state ownership. The local supply of spring-fall range exceeds the supply of summer range, so that the ownership of summer range is the land base for lease and control of spring-fall range. Land tenure and control is a minor management problem for this ranch since ownership of summer range is the basis for control of an adequate supply of spring-fall range.

Ranch and Range Improvements. The improvements are adequate, except that more and better planned holding pens are needed adjacent to the lambing sheds to facilitate making up the ewe and lamb bands during lambing, and there is some need for additional water development on the summer range.

The Balance of Range and Cropland Capacities. The spring-fall range and feed-crop capacity of this ranch overbalance the capacity of the summer range. As a result, the summer range is over-used and the trend of the condition of this range is down. Partly as a consequence of this, the lamb weights are considerably below what could be attained. This lack of balance in land capacities should be the key to management planning for this ranch. The number of sheep should be reduced to the sustainable capacity of the summer range—about 6000 head—and the acreage of the spring-fall range under leases decreased accordingly.

Livestock Management. Wool production has been somewhat overemphasized in the program of this ranch, considering its pos-sibilities for lamb production. In an attempt to get longer staple and somewhat coarser wool, the operator has made considerable use of crossbred (Rambouillet x Corriedale) sires with Rambouil-let ewes in breeding for replacements. The lamb crop has averaged about 70 percent, with an average weight of sixty pounds. This ranch has used an inadequate number of bucks—an average of one to 55 ewes. This ranch should try operating a band of cross-bred whitefaced ewes and if they prove adaptable, shift to that type of sheep. With a reduced number of ewes, the feed-crop pro-duction would permit a better level of winter feeding than has been practiced (325 pounds per ewe for a four-month season) and the summer range, when used nearer its capacity, should enable this ranch to achieve a much higher market turnoff of lambs.

Kind of Livestock. Range and crop feeds of this ranch are well suited to lamb production. The ranch operates a supplemental cattle enterprise of about 100 head. It might eventually be desir-able to expand this somewhat, depending on the results of the pro-posed change in sheep numbers and land acreages to get a better balance in the range and feed-crop capacities of the ranch.

Markets and Marketing Methods. The practice has been to sell the lambs to country buyers. If this ranch makes the changes to place more emphasis upon lamb production, and, as appears prob-able, a fair percentage of slaughter-class lambs can be produced, the operator should test the results of shipment to central markets or sale through the local auction market.

Operating Costs. These proposed changes will increase the oper-ating costs of this ranch but the indications are that such changes will also increase the gross revenue by two to three times the amount of the increase in operation costs. The annual operating costs, before the wartime inflation, averaged about $3.50 per ewe, or the level of the plains sheep ranch rather than that of the mountain valley ranch that has good lamb production opportunities. The labor input of this ranch has averaged one man-year of labor time per 700 ewes—comparable with the plains sheep ranch operation that does not have the resource opportunity for high lamb production.

Ranch Investment Values and Finance. This ranch now carries a totaled deeded land capitalization of $14 per ewe for the present number of 8000 head. The present system of management has yielded a return of about 1 percent on this investment, after allowance of an operator wage of $.60 per ewe and a 5 percent return on a normal value for the sheep. This ranch should return 5 percent on a $12 land investment per ewe, after making such allowance as the above for operator wage and interest in livestock values. The present indebtedness of the ranch is not excessive— $3.35 per head on the sheep, and $3.10 per-head land debt, based on the present number of 8000 ewes.

CHAPTER VIII

The Use of Federal Public Lands

Probably something like half the stock ranches of the eleven western states make some use of one or more of the several kinds of federal public lands. Laws and regulations governing the grazing use of these public lands differ considerably. Our purpose here is to describe the important features of federal public land administration in the West as related to the use of these lands for grazing. We will also describe what appear to be the noteworthy interrelationships between western federal public land administration and the management of the western stock ranches that use these lands.

In qualification of the materials in the following pages it should be noted that the information here presented on public land legislation and programs applies to the *present*. Such legislation, programs, and administration will evolve and may undergo rapid change, as illustrated by the recent reorganization of the Taylor Act administration.

Materials in this chapter will be limited to the federal public lands. Grazing use of the state-grant lands of the eleven western states is a significant item. Too many differences prevail in the handling of these lands by the states, however, for any comprehensive description in this chapter of the administration of these lands. Most of the eleven western states received a grant of two sections (sections 16 and 36) per township. Some of these states received four sections per township as their grant. These lands are subject to sale, but at a minimum price, as defined by the grant statute, too high for normal sale as grazing land. These lands are leased from the states by stock ranches and farms for grazing and farming. Many differences prevail among the states in their provisions for lease prices, term of years leased, acreages leased to any one individual, regulation of use, and the like.

KINDS AND IMPORTANCE OF FEDERAL PUBLIC LANDS

Federal public lands in the eleven western states account for approximately 30 percent of the present grazing use of rangelands in these states. These lands are, however, largely seasonal in use, and as a consequence, more than 30 percent of the ranches use these lands — as previously stated, probably about 50 percent of the total number of stock ranches.

The estimated present grazing use of the public lands is given, in round numbers, in the tabulation. These figures are in animal unit months (one head of cattle or five ewes for one month). The

	Number
Federal grazing district lands	15,000,000
National forests	8,500,000
Indian reservations	7,500,000
General land office lands	2,225,000
Land utilization project lands	1,300,000

Indian reservation lands are not in reality public lands. Since, however, the policies in the use of these lands are to a considerable degree determined by a federal agency, the U.S. Indian Service, we shall consider them as federally administered lands.

In addition to the grazing use of the federal lands listed above, there is some grazing by domestic livestock on the wildlife refuges and game ranges administered by the U.S. Fish and Wildlife Service. If the lands withdrawn for game ranges lie within the boundaries of federal grazing districts organized under the provisions of the Taylor Act, the grazing administration is handled by the U.S. Department of the Interior's Bureau of Land Management (now the administrative agency for the Taylor Act). The Charles Sheldon game range of northern Nevada and the Hart Mountain game range of southern Oregon are illustrations of this arrangement. Other withdrawals from the public domain of some local and limited importance for grazing are the reclamation site withdrawals, and the stock driveway withdrawals made before the passage of the Taylor Act. Where these withdrawals lie within the boundaries of grazing districts organized under the Taylor Act the grazing is administered by the Bureau of Land Management.

FEDERAL GRAZING DISTRICT LANDS

The Status Patterns. When the act of Congress known as the Taylor Grazing Act was passed in 1934, some 170 million acres of unreserved and unappropriated public domain lands remained in the eleven western states. Nevada, Arizona, and Utah had the largest acreages. These lands were the "leftovers" from the homestead selections and from the reservations of public domain lands for national forests, national parks, and permanent public ownership of other lands with high public values. This does not mean that all were "poor" lands for grazing use, however; often homesteading of the lands with the water and the lands in key locations gave fairly good control over the grazing use of the nearby public domain. As a result the ranches located in large areas of public domain lands did not feel any urgent need for ownership of more than the lands with water and key locations.

At the time the Taylor Act was passed there were, as a result of the past homestead selections, several distinct status patterns for the remaining public domain lands. One of these patterns characterizes the remaining public domain lands of the northern plains — eastern Montana and Wyoming. This is a scattered pattern, varying from isolated tracts to the "patchy" areas of public domain lands in badlands, along stream breaks on alkali soils, and the like. Another pattern typifies the intermediate and higher lands of the intermountain region — the sagebrush lands and the piñon-juniper lands. These lands have been heavily selected and patented for grazing and other uses. From 30 to 60 percent of these lands (the best of them) are now under private ownership. A third pattern may be seen in the low desert lands, the sheep winter ranges and the yearlong desert cattle ranges. For these lands private ownership is limited to the lands with the water and the key locations for control of the grazing on the public domain.

The Taylor Act. Through private ownership of key lands within the public domain lands and leasing railroad-grant lands and state-grant lands, the ranches using the public domain lands had attained some measure of control and management of these lands before the Taylor Act was passed. Even so, the public domain grazing was largely "open" and unregulated in season, numbers, and kind of stock.

The Taylor Grazing Act was enacted on June 28, 1934. This act, with subsequent amendments, provides for the inclusion of a maximum of 142 million acres of the remaining unreserved and unappropriated public domain lands of the United States (exclusive of Alaska) in grazing districts, for a plan of administration of such lands for grazing use, and for administrative organization for management. To date, of the approximately 170 million acres of unreserved and unappropriated public domain lands remaining when the act was passed, 133,420,000 acres are included in the sixty grazing districts that have been organized under the act.

Subsequent to the enactment of this law, the remaining public domain lands were withdrawn from homestead entry under the homestead laws by executive order on November 26, 1934. However, the Taylor Act provides that the Secretary of the Interior may examine and classify such lands, both outside and inside grazing districts, for possible entry under the homestead acts, but such homestead entry cannot be in tracts exceeding 320 acres. The Taylor Act does not limit location and entry of mining claims under the provisions of the mining laws.

Essentially these are the provisions of the Taylor Act for administration of the public domain lands for grazing: (1) creation of districts for the administration and management of these lands (Figure 3 shows the organization of such districts to date); (2) promulgation of regulations, consistent with the act, for administration and for the establishment of an organization to handle the administration of the act; (3) issuance of permits to grazing users; (4) making of improvements; (5) sale and lease to users of isolated tracts of public domain outside the boundaries of grazing districts; and (6) organization of advisory boards for each of the grazing districts.

An administrative organization known as the Grazing Service was set up in the U.S. Department of the Interior soon after the Taylor Act was passed. The staff of this service consisted of a director and assistants, a regional officer in each state, a district officer in charge of each district, and technicians assigned to work with the districts. Recently this organization has been merged with the General Land Office of the U.S. Department of the Interior to form the Bureau of Land Management, which handles the administration of the districts and the leasing and sale of lands outside of districts.

The administrative rules and regulations which have been promulgated under the provisions of the Taylor Act are embodied in

FIGURE 3. GRAZING DISTRICTS UNDER THE TAYLOR GRAZING ACT, DEPARTMENT OF THE INTERIOR, BUREAU OF LAND MANAGEMENT

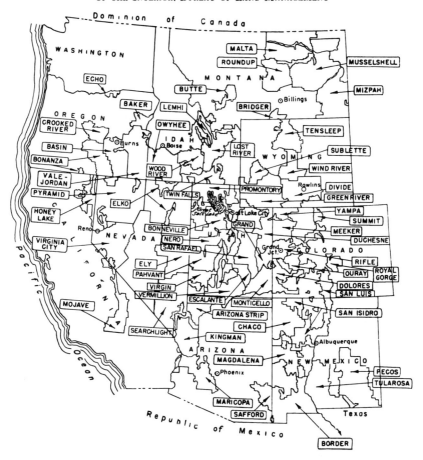

a publication known as *The Federal Range Code*. These rules and regulations were developed by the Grazing Service and the advisory boards, and when approved by the Secretary of the Interior, they became a body of administrative law within the provisions and framework of the Taylor Act.

To the stock ranch operator, the most important features of the Taylor Act and of *The Federal Range Code* are those relating to the issuance of grazing permits; that is, the basis for granting permits, the tenure of the permits, their tranferability, and so on. Section 3 of the Taylor Act provides that ". . . preference shall be given in the issuance of permits to those within or near a district who are land owners engaged in the livestock business, bona fide occupants or settlers, or owners of water or water rights, as may be necessary to permit the proper use of lands, water or water rights owned, occupied, or leased by them. . . ." In carrying out this provision of the act, the *Code* sets up certain rules about the requirements for the base property of the ranch and for classification of such base property to establish the relative order of preference of the applications for grazing permits on the public land.

The concept of base property — the owned or leased feed-crop land, seasonal rangeland, or range livestock water — recognizes the seasonal use interdependency of the ranch land and the public land of the grazing district. Administrative rules regarding the amount and kinds of base property that the ranches should have and the season of use of such base property are adapted to the recent operating methods of the ranches and the land economy of the district. Item 1 of subsection c of section 6 of the *Code* provides that "No license or permit will be issued to any applicant unless he is able to show that he possesses adequate feed to support his licensed or permitted livestock during the period of time for which they are to be off the Federal range." Thus, the intent of the rule regarding base property is to fit the use of the public land into the normal year-round cycle of rangeland capacity and feed-crop production of the ranches.

Code provisions state that the Class One base property for qualifying for a grazing permit is the land or water that has established its use dependency on the public lands through past use history in livestock production. The period of use for meeting this qualification is stated in the *Code* to be two consecutive years or any three years of the five-year period immediately preceding June 28, 1934 (except that for districts organized after June 28, 1938, the prior use is two consecutive years or any three years within the five-year period immediately preceding the organization of the district).

Further *Code* provisions state that the dependency of base property by use during the base period must be established through permit application within one year after creation of the grazing district in which the public lands which were used are located. The purpose of this provision is to facilitate the issuance of term permits to the qualified applicants.

Class Two base property for grazing permits is land or water that is dependent by location. That is, a land or water property base that is so located and of such character as to have an economic use interdependency with the public land range, but that does not qualify by use history during the base period, is of second priority and will qualify for a permit only when there is unused capacity of grazing district lands after all the permit applications of the Class One base properties are cleared.

Thus, while the administrative concept of base property for grazing permits is a land use and ranch economy concept, it is possible under the provisions of the *Code* that, through past accidents of use history, a superior base property is denied a permit and an inferior base property is granted one. However, from a practical administrative standpoint it was desirable and necessary to place the first emphasis in the granting of permits on the recent use history of the lands. Eventually the grazing permits will shift to the most productive of the base properties. Paragraph b of section 7 of the *Code* provides, with certain qualifications, for the transfer of a permit from one base property to another, and such a provision is desirable in permitting the economic adjustments between base properties and grazing permits.

By provision of the Taylor Act, term grazing permits may be issued for a maximum period of ten years. During the first years of the administration of the act, annual permits or "licenses" were issued, but approximately 65 percent of the 23,000 permits are now on a term basis. The *Code* states that there shall be a proportional reduction in the grazing permit when the holder of such permit ceases to make any substantial use of his base property in connection with his year-round livestock operation. It is also provided in the *Code* that there shall be a proportional reduction in

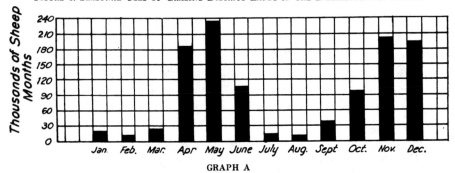

GRAPH A

Months of permitted use by sheep on the Lost River Grazing District of Idaho in 1940. This spring-fall season of use by sheep is typical for the sagebrush-grass range type of southern Idaho, eastern Oregon, northern Nevada, northern Utah, and western Colorado.

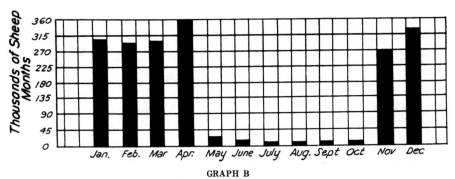

GRAPH B

Months of permitted use by sheep on the Bonneville Grazing District of Utah in 1940. This seasonal pattern of use by sheep is characteristic for the western desert-shrub type of western Utah, eastern Nevada, and southwestern Wyoming. The principal use of this type is winter sheep range.

FIGURE 4 continued.

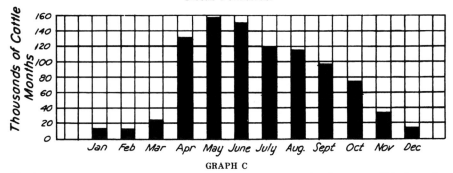

GRAPH C

Months of permitted use by cattle on the Elko Grazing District of Nevada in 1940. This shows the usual season of use by cattle of the sagebrush-grass ranges of northern Nevada, eastern Oregon, northern Utah, and southern Idaho. The cattle are maintained on hay from December through March and are moved from the grazing district range to ranch pastures and hay and crop aftermath starting in July.

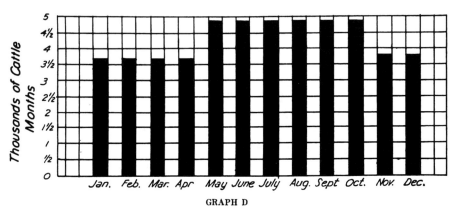

GRAPH D

Months of permitted use by cattle on the Searchlight Grazing District of Nevada in 1940. This shows the nearly yearlong season of cattle use of the semidesert grass and browse type of southern Nevada and Utah and the southwestern region grazing districts. The principal deeded land base is water development.

SOURCE: "A Graphic Summary of Grazing on the Public Lands of the Intermountain Region," mimeographed publication of the Intermountain Forest and Range Experiment Station, U.S. Forest Service, Ogden, Utah. June 15, 1941.

217

the permit if at any time during the life of the permit the holder of the permit loses control of all or part of his base property and does not within a reasonable time obtain control of other base property.

At present the *Code* provides charges for grazing on the grazing district lands of 8 cents a head per month for cattle and 1⅗ cents a head per month for sheep, for animals over six months of age. This is a flat rate that is applied uniformly to all the districts. Although this is now a uniform rate, and a very low one, it is probable that eventually a plan of charges will be developed that recognizes variations in the quality and value of the lands, and that places the general level of charges more in line with competitive market rates for lands similar to the lands of the grazing districts.

Seasons of Use Established for Grazing District Lands. Sheep grazing seasons established on the grazing district lands of the intermountain region are principally spring-fall and winter. The season of sheep use is mainly spring-fall for the districts of southern Idaho, eastern Oregon, and western Colorado, where the sheep ranges are principally the sagebrush-grass type. This season of use is illustrated by Graph A of Figure 4. Base property for this spring-fall season of use is generally feed-crop lands. Usually the season of sheep use on the desert shrub ranges of the districts of western Wyoming, western Utah, and eastern Nevada is the winter months, as shown by Graph B of Figure 4. Base property for this use is spring-fall or summer range. There is some summer use by both sheep and cattle on the higher lands of the grazing districts of the intermountain region.

A typical season of use for cattle on the sagebrush-grass ranges of the grazing districts in the northern parts of the intermountain region is illustrated by Graph C of Figure 4. This shows how the cattle are moved onto the grazing district land in April and are moved off starting in midsummer, as the early-growing annuals dry up and water becomes scarce in the lower country. These cattle are moved onto the fenced rangelands of the ranches and the hay meadows in late summer and early fall. Ranches with national forest permits move the cattle from the grazing district in May. The large acreage sagebrush-grass ranges of these grazing districts once afforded a longer season of summer use by cattle and they will do so again, eventually, if management can restore a substantial complement of the perennial bunchgrasses and if adapted introduced grasses can be seeded as economically as now appears possible.

Cattle use on the semidesert grassland and desert shrub ranges of the grazing districts in the southern part of the intermountain region is to a large degree yearlong, with range water on limited private land ownership as the principal base property of the ranches. This is illustrated by Graph D of Figure 4. This yearlong season of use by cattle is also characteristic of the semidesert grasslands and browse ranges at the higher elevations of the grazing districts in the southwestern region.

District Management Plans. In each of the grazing districts administrative subdivisions or "units" (based upon differences in season of use and kind of stock) have been set up, migration routes established, and individual (or group) allotments mapped out for the users. The management and improvement program for the individual range allotments has not, in the general picture, proceeded far as yet. This is important, if the administration of the Taylor Act is to be fully effective. Some such management plans for range allotments have been prepared and put into effect. One such plan is illustrated in Figure 5.

A Look at the Future. Because of a complex set of circumstances that need not be described here, the future program of administration under the Taylor Act is uncertain. Granting of the public domain lands to the states, sale of these lands to the present grazing permittees, and other substitute measures are being advocated. Questions are being raised about whether the Taylor Act is a suitable law for handling these lands, and whether serious errors of policy have not been made in the administration of the act.

Probably the administration of the act erred in extending the organization of districts too far. Probably the scattered public domain lands in the central and eastern Montana districts should have been left to the General Land Office for sale and lease under sections 14 and 15 of the Taylor Act. (For illustration, there are approximately one million acres of public domain land out of a total of 9,100,000 acres within the boundaries of the Musselshell district in central Montana.) This observation probably applies also to some of the districts in other states.

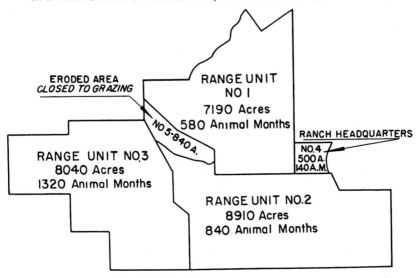

FIGURE 5. AN ILLUSTRATION OF MANAGEMENT PLANNING FOR A CATTLE ALLOTMENT
ON FEDERAL GRAZING DISTRICT LANDS (MAGDALENA DISTRICT, NEW MEXICO)

1. Winter range. Rough topography provides natural shelter. Browse and bunchgrass range feed. General elevation about 6500. This unit, with the ranch headquarters, is used for the three months from November through January.

2. Spring range. Used from February 1 to May 15. General elevation about 5000. Early weed and grass feed.

3. Summer and fall range. Used from May 15 to November 1. A grama grass range. General elevation about 6000.

This allotment has a capacity for 240 head of cattle. It is necessary to feed some range supplements on the winter range to attain a balanced seasonal relationship for the management units of the allotment. Before the organization of the grazing district the rangelands of this allotment were stocked with some 400 cattle, 30 horses, and 900 sheep throughout the year, with very poor production results.

After the district was organized but before the preparation of the management plan, the cattle drifted at will over the entire allotment. The result was a heavy use, throughout the year, of what is now range unit number 1 and an underuse of the number 3 unit. This was due at least in part to a lack of water development on number 3, the unit which is naturally suited for the summer and fall range. There are now six water developments on this unit. This is ample.

The ranch headquarters are owned by the user of this allotment who also owns and leases 7100 acres of private land and state land interspersed with the 17,000 acres of federal public land in this allotment. This ownership situation is fairly characteristic of the cattle allotments of the federal grazing districts and emphasizes the necessity for joint efforts between the administrative agency and the range user in the development and application of management plans.

220

It seems that the act might well have been made more flexible by providing for some outright leasing to individual users or to associations of users within organized districts where there is an intermingling of the public lands and the private lands owned and leased by the grazing users. Such leases of the public domain in districts or parts of districts with intermingled public and private lands could carry the desirable stipulations concerning season of use, numbers of livestock, and so on. This ownership pattern characterizes much of the sagebrush and piñon-juniper lands of the intermountain region, intermediate between the national forests and the desert lands. These intermediate lands and some of the lower desert lands are yielding much, perhaps most, of the silt now going into valuable irrigation and power reservoirs. Such lands need conservation measures — *badly*. We may see them taken out of grazing use if the present silt yield from these lands, both public and private, continues.

Provision for administrative and technical personnel to handle administration under the Taylor Act has been strangely inadequate. Public domain lands of 7 million acres in one grazing district, which has in all some 11 million acres of gross area, are "handled" by a grazier and assistant.

NATIONAL FOREST LANDS

The Enabling Legislation. Congress passed the original act providing for the reservation of national forest lands from the public domain in 1891. On June 4, 1897, Congress passed an act providing for the administration of national forest lands. The purposes of the legislation are set forth in this act of 1897 in general terms. The Secretary of the Interior (changed to Secretary of Agriculture by the legislative act of February 1, 1905) is directed to make such rules and regulations and establish such service as will insure the purposes of such reservations, ". . . namely, to regulate their occupancy and use and to preserve the forests thereon . . ." It appears worthy of note that the timber and watershed uses are specifically mentioned in this act providing for administration, but that grazing use is not so mentioned.

Streamflow protection and timber production represent, as a general rule, the highest values in the management of the national forest lands, and this is recognized in the legislation and in the

Ranch properties with a high dependency on mountain range sometimes conflict with the watershed values of the mountain lands.
(Neil W. Johnson Photo)

policies and regulations that have been developed to promote the intent of the legislation. It should be noted, however, that the use of the national forests for the production of domestic livestock has been clearly recognized and established since the passage of the act of February 1, 1905, transferring the administration of these lands from the Department of the Interior to the Department of Agriculture.

Administrative Policies. All the policies developed to carry out the objectives of the legislation for administration of the national forest lands have stressed the concept of "multiple land use" for such lands. This concept is a recognition of the several values of the national forest lands — watersheds, timber production, recreation, grazing, and wildlife — and of the fact that two or more of these values are generally coexistent on the same lands. Highest use values for these lands are to be attained, consequently, by what appears to be the best combination of uses, and certain uses may reach far beyond the local community economy. Watershed and recreational uses are illustrations.

Regulations developed by authority of the act of 1897 to express policies for the use and management of national forest lands for grazing are contained in Volume III of the U.S. Department of Agriculture Forest Service Manual, "U.S. Forest Service Manual for the National Forest Protection and Management of the National Forests." This manual is a compendium of national forest management regulations, together with a considerable amount of technical information and instructions for the use of trained Forest Service personnel in the management of national forest lands. These regulations for national forest administration are prepared in codified form and published in the *Federal Register*. The regulations as they appear in the manual are available to anyone in the offices of the district ranger or the forest supervisor.

Administrative Organization. The national forest protection and management branch or division of the U.S. Forest Service handles the administration of national forest lands. National headquarters of the Forest Service are located in the U.S. Department of Agriculture, Washington, D.C. An assistant chief responsible to the chief of the Forest Service is in charge of the branch of national forest administration, which includes the division of range management. The division of range management is headed by a division chief who is responsible to the assistant chief for all range management work on the national forests.

Administration of the national forest lands is handled through regional offices with a regional forester in charge. There are six regions in the eleven western states (see Figure 6), the regional offices of which are located at Missoula, Montana; Denver, Colorado; Albuquerque, New Mexico; Ogden, Utah; San Francisco, California; and Portland, Oregon.

A research branch of the U.S. Forest Service, with an assistant chief and staff in the Washington office, operates a forest and range experiment station in each of the six western regions. An important feature of the research branch is a division of range research in the Washington office and in each of the western forest and range experiment stations. This branch of the Forest Service is designated as the research agency having responsibility for all federal government scientific investigations in the field of range management, both for public lands and for privately owned lands. Forest and range experiment station headquarters in the western regions are located at Missoula, Montana; Fort Collins, Colorado;

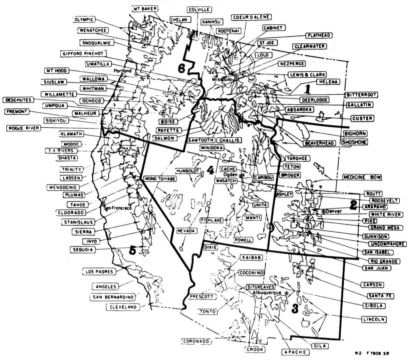

Tucson, Arizona; Ogden, Utah; Berkeley, California; and Portland, Oregon.

The national forest is the administrative unit for the management of national forest lands. There are at present 104 national forests in the eleven western states, with a forest supervisor as the administrative officer in charge of each forest. Each national forest is divided into ranger districts, and a district ranger, responsible to the supervisor, is the administrative officer of the district. The forest supervisor and district rangers are general administrative officers who must be qualified to handle the several different phases of national forest management. The regional office provides policy guidance and technical service for all the national forest offices of the region. Regional office staffs are organized on functional lines for service to the forest offices. Each of the western regional offices has a division of range management, staffed by specialists in this field. These specialists give the forest offices technical service and policy guidance in the management of national forest lands for grazing use. The regional forester correlates this service and the policy work of the regional office specialists in range management, timber management, watersheds, wildlife, recreation, and the like.

The Range Management Unit. A grazing allotment constitutes the geographic division upon which national forest range management is based. Range allotment boundaries are posted, fenced, bounded by a natural barrier, or otherwise defined. The sheep allotment is usually an individual allotment to a user; the cattle allotments are frequently used by a group of permittees who function as an association for cooperation with the Forest Service in the use and management of the allotment. National forest range allotments are divided into management units or "camp units," and the management plan for the allotment shows the plan of movement of livestock over the allotment, the opening and closing dates of use, the band days or animal days of use in each unit, the use of salt and bed grounds, the plans for future range development and improvements, and the like (see Figure 2). When a cattle allotment is used by a group of permittees, such a group may agree on certain livestock management features, such as the nummer of riders, number and kind of bulls, cooperative work with the Forest Service on improvements, and so on.

Seasonal Grazing Use and the Ranch Property Base. Some 77 million acres of the 136 million acres of national forest lands in the eleven western states are now used for the grazing of domestic livestock.* This land (except in the southwestern region) is principally summer range, and its use is, consequently, highly interdependent with other seasonal ranges and feed-crop production. The logical base property of the ranch for the use of the national forest grazing is, consequently, feed-crop land and spring-fall range, with the emphasis on the latter where the winter feeding season is short. The typical seasonal use pattern of national forest lands by cattle and sheep is illustrated by Figure 7. There is, however, some yearlong grazing on national forests in Arizona, New Mexico, and southern California.

The typical season of use of national forest summer range, as il-

* Most of the land not so used is not suited to grazing use. Several million acres are closed to use for city watershed purposes.

FIGURE 7. SEASONAL GRAZING USE OF THE NATIONAL FOREST LANDS OF THE
ROCKY MOUNTAIN REGION

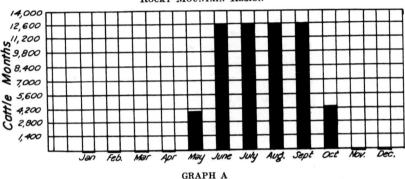

GRAPH A

Months of permitted use by cattle on the Caribou National Forest of Idaho in 1939.
This shows the typical seasonal pattern of use by cattle of the national forest lands of
the Rocky Mountain region.

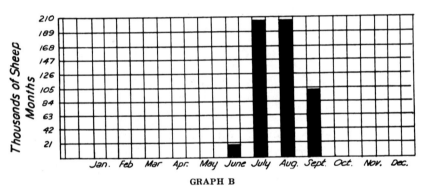

GRAPH B

Months of permitted use by sheep on the Caribou National Forest of Idaho in 1939.
This typifies the season of use by sheep of the national forests in the Rocky Mountain
region. The sheep enter the national forests later than the cattle and move to the
higher ranges for a period of 2½ to 3 months of grazing.

SOURCE: "A Graphic Summary of Grazing on the Public Lands of the Intermountain
Region." mimeographed publication of the Intermountain Forest and Range Experiment Station, U.S. Forest Service, Ogden, Utah. June 15, 1941.

FIGURE 8. LOCATION OF SHEEP AND CATTLE RANGES IN THE
BIGHORN NATIONAL FOREST

This shows the typical pattern of use of national forest summer range in the
Rocky Mountain region. The high country is used principally as sheep range;
the lower elevations around the mountain area are used by cattle.

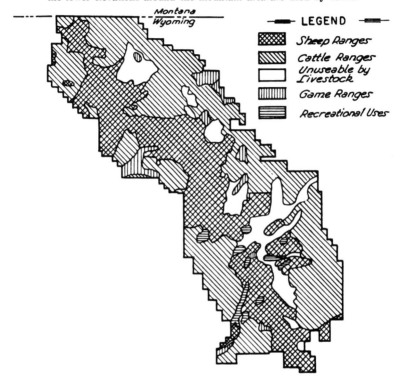

lustrated by the Caribou National Forest, is considerably shorter
for sheep than for cattle. This contrast between sheep and cattle
in season of use occurs because the sheep use the highest ranges,
where the weed and browse feeds are more prevalent, but the cattle use the lower country where the bunchgrasses are more prevalent. This typical use of different elevations of mountain grazing
land by cattle and sheep is further illustrated by Figure 8. This
shows the location of sheep and cattle ranges on the Bighorn National Forest (north central Wyoming). Sheep are taken into the
sheep allotments as rapidly as feasible over the trail routes when
the season opens. Cattle use is a controlled seasonal drift up and
back down the cattle allotments on the mountain slopes and lower
elevations.

Grazing Permittees. The present plan of use and administrative regulations for grazing on the national forest lands have been evolved over a period of approximately forty years. Many of these lands were grazed as open public domain before they were withdrawn for national forest administration. Temporary grazing permits were issued to the users at the time a national forest was placed under administration. Soon thereafter an approach based on land use and ranch and farm economy was developed as the basis for issuing grazing permits to the operators of farms and ranches.

Since the upland summer range of the western states is generally limited in supply in relation to other seasonal ranges and feed crops, the summer range of the national forests has always had a high demand. This condition makes it necessary for the Forest Service, as the administrative agency, to make a choice and selection among the range livestock producers who wish to use national forest lands.

The alternative to making a selection from among the different possible users would be competitive bid leasing of national forest grazing, but it is believed that this procedure would likely be inimical to good management of the several use values of national forest resources. These resource relationships and the necessity for choice among the possible users are basic in the plan of grazing use that has been developed as a part of the program of national forest management.

Ownership of ranch property that has a high economic interdependency with the grazing use of national forest land constitutes the primary requirement for holding a national forest grazing permit. This means ownership of feed-crop land that is limited in other agricultural alternatives, or of seasonal rangeland that is naturally complementary to the season on the uplands, and a ranch location that gives efficient access to the upland grazing. The Forest Service has generally required ownership rather than lease of the base property, as favoring stability of users and, consequently, stability of management of the national forest grazing.

The policy in selecting users has also favored the family-size unit, again because of the factor of stability. In furthering such a policy, limits are set for the lower and upper size of permits, the lower limit being the size of operation that will provide a family living, and the upper limit being the size that will achieve good technical operating efficiency for the livestock enterprise. It is, of course, necessary to fit to the local economy the regulated limits in the size of permits, and there are, consequently, considerable variations in the size of the upper and lower limits for different forests. There are some individual cases of permits recognized for a special limit above the upper limit, but no guarantee is made that these permits will not be reduced to the upper limit for purposes of distribution to permittees whose permits are below the upper limit, or to new applicants who can qualify as national forest grazing users.

Preference and Temporary Permits. The 22,000 paid* grazing permits that are now in effect on the national forests of the six western Forest Service regions are classed as "preference" or "temporary." Preference permits are those that have been established through past use in connection with qualified base property. Temporary permits are issued for only one year at a time, when additional range becomes available (because of relinquishment, cancellations, and reductions of existing permits; granting of nonuse to preference holders; or an improvement in range conditions) and they may or may not be renewed the succeeding year. Temporary permits may be issued to new applicants or to preference holders. When a temporary permit has been in effect for five consecutive years, it may under certain conditions be recognized as a preference.

Preference permits are generally on a ten-year term basis and are construed as term agreements, within their provisions, between the permittees and the federal government. No commitment is made for the numbers in the permits above the upper limit, as stated before. The Forest Service may renew the preference to the purchaser of the ranch property or the livestock, either one or both, of a preference holder. This sale must be a bona fide transaction, and the purchaser of livestock only must qualify with ownership of other base property. There are now, consequently, two methods (other than inheritance) for the new user to acquire a grazing preference on national forest lands: (1) through a waiver of an established preference by the holder to the government

---

* Free use permits are generally granted for up to ten head of animals kept primarily for home use or for work stock.

and subsequent renewal to the purchaser of ranch property and/or livestock; and (2) by obtaining a temporary permit which, after the required period of continuous use for dependent ranch property, may under certain conditions be recognized as a preference.

Any permit may be reduced at any time for protection of the resource, as may be needed, but such protection cuts are to be restored to the preference permits when and if the range condition warrants, before any temporary permits might be granted to any other users of the allotment.

A reduction of 10 percent may be made in a preference when it is renewed to the buyer of the ranch property and the livestock of a preference holder. A reduction of 20 percent may be made if only the livestock or only the land is purchased. The reduction may be used to restore protection cuts to other preference holders, to grant new temporary permits to present permittees or to new permittees, or to permanently decrease the use of the range for resource protection. A preference holder may be given nonuse of his permit for a period of three years if he does not wish to use national forest range, and he may, by agreement, take nonuse for part or all of a preference for a period up to five years when it is decided that a temporary reduction in use is necessary to protect the resource.

Advisory Boards. Grazing advisory boards, composed of elected representatives of users, have been an important feature of the management of national forest grazing since the forest lands were placed under administration. These local advisory boards are organized on the basis of their community of interest as a group of users in the grazing use and management of an area of national forest land. Local advisory boards select the members who are to represent the local group in the forest-wide advisory board, and this board works with the national forest administration in matters of over-all policy in the use and management of the national forest lands for grazing.

National forest grazing advisory boards are not given legal status as a part of the U.S. Forest Service. They function in an advisory capacity only and have no administrative authority. The organization and use of these boards has been a matter of Forest Service policy under the provisions of the Organic Administrative Act of 1897, which directs the administrative agency to establish such services as will insure the objectives of the acts under which the national forests are created.

Grazing Fees. Charges for grazing on the national forest lands are based upon an appraisal of the differences in the range quality, location, accessibility, improvements, and other factors affecting the value of these lands for grazing. Grazing fees vary considerably between regions, forests, and ranger districts. Base rates for the grazing fees now prevailing were established in 1931. A variation is made above and below these base rates in accordance with the annual variations in the average market prices (for the eleven western states) received for livestock. For example, the base rate for cattle grazing on one national forest is $.20 per animal month, and the average price received by growers, for cattle, during the period of the appraisal when this base rate was determined was a western average of $6.60 per hundredweight. If the average price received by the western growers is estimated at $7.25 in 1938, or 10 percent above the appraisal base period price, the grazing fee for 1939 is $.22 per animal month, or 10 percent above the base fee.

As a general average the base rate for cattle on the national forests of the eleven western states is $.14½ per animal month, and the western average base price for cattle associated with this base rate was determined to be $6.60 per hundredweight. The comparable figures for sheep are a base grazing fee of $.04½ per ewe month, and a western average lamb price of $9.15. There is considerable variation above and below these average western base rate grazing fees for individual forests and ranger districts. Most of the base rates for cattle are between $.10 and $.20 per animal month, and most of the base rates for sheep are between $.02½ and $.06 per animal month. These base rates for national forest grazing are considerably below competitive market values. The base price for the sliding scale of fees above and below the base fee rates is high for lambs, compared with the base price for cattle. Consequently, the sheep grazing fees are above the base fees only during high prices.

The principal appraisal factors considered in determining the differences in the base value of national forest grazing were: (1) the quality of the range as influenced by the predominance of good range plant species and the density of plant cover on the ground; (2) the adequacy of water and water facilities for avoiding excessive trail movement of the livestock to water; (3) the adequacy of improvements for the management of livestock on the range; (4) the location of the range for moderate trail movement from the base property to the range, and from the range to shipping points; and (5) the accessibility of the range by road and trail for easy movement of the livestock and of camp supplies into the range.

The first three of these factors have an important influence on livestock production gains and market weights, and apply about the same to any rangeland, whether it is privately owned fenced land or upland "wild" land. The last two of these appraisal factors take into account the added costs of using lands that are remote from the ranch, and these factors are applicable especially to the national forest and federal grazing district lands.

Rental prices paid for comparable private lands were used as the basis for the appraisal of national forest grazing. Some compromises were made in the appraisal results in the grazing fees that were established.

INDIAN RESERVATION LANDS

Leases Rather Than Grazing Permits. Grazing of livestock on Indian reservation lands by the operators of adjacent stock ranches is important to the ranch economy of some localities. The Blackfoot and Bighorn reservations in Montana, the Fort Hall in Idaho, and the Pine Ridge and Rosebud in South Dakota are illustrations. Most of this grazing use by the adjacent and nearby ranches is on the "trust patent" lands but there is also considerable leasing of the tribal lands of the reservations.

Grazing use of these Indian reservation lands is secured through competitive-bid lease. Leasing is handled through the U.S. Indian Service agency office for the reservation, in cooperation with the tribal council for the tribal lands. There is also in some cases considerable leasing of grazing lands from the Indian owners who have acquired final ownership of their allotments of reservation lands. The trust patents on the individual allotments are held by the Indian agency office for a period of twenty-five years and the leasing of the trust patent lands is handled by the Indian agency office for the individual Indian allottees.

Use Seasons and Property Bases. The several northern Indian reservations contain some sizable areas of good plains and foothill grazing lands. This is particularly true of the Indian reservations of Montana, Wyoming, South Dakota, and Idaho. Grazing use of these lands is for the most part season-long—that is, for the whole year except the period of winter feeding. Ranch lands needed to use Indian reservation grazing are, consequently, primarily haylands and other feed-crop lands. Grazing use of Indian reservation lands by stock ranches is principally in the four states mentioned above. Grazing use of the large reservations of New Mexico and Arizona is largely by Indian livestock.

Lease Costs. Costs of using the Indian reservation lands generally run considerably higher than those for the national forests and the grazing districts. Unless term leases can be obtained, the Indian reservation leases are less secure and stable than are the public land permits. Costs of leases on the Indian reservation lands have in the past ranged from $.20 to $.75 per animal month for cattle, and from $.05 to $.20 per animal month for sheep. Since 1945 the cattle leases on these lands have risen as high as $1.50 per animal month.

Types of Ranch Users. It is rather generally true that the policies in granting grazing leases on the Indian reservations have favored the large operators who are able to move livestock considerable distances and handle the leases in large blocks. A limiting factor in the use of Indian reservation grazing in the northern states is the frequently insufficient development of croplands and of winter feed production within or near the reservation lands. As a consequence, some of the large operations using Indian land leases have attempted yearlong grazing with inadequate reserves of winter feed to meet the emergency requirements. Such operations have often been hazardous and semispeculative in character.

LAND UTILIZATION PROJECT LANDS

The Repurchase Program. Beginning in 1934 a federal government program was initiated for the re-acquirement of lands in certain areas that had been settled for dry-land crop agriculture and were judged to be clearly unsuited to such use. In this program certain areas known as project areas were selected, and within such areas the poorer and smaller farm units were acquired, if possible. Most of the activity under this program was in the northern plains — eastern Montana and Wyoming and western North and South Dakota. Usually the policy in this program was to purchase a minor percentage of the land of a sizable area rather than to attempt to buy large, solid blocks of land. Small farm units were bought from the operators, homestead lands from absentee owners, and some tax-deed land from the counties. The purpose of this pattern of acquisition was to reseed and otherwise restore the acquired land to grazing and pasturage and fit its use into the operations of the remaining farm and ranch units, to readjust such units to a ranch or stock farm type of operation.

Cooperative Grazing Association Districts. This land utilization program is a small one in the total picture of western stock ranch operations. The feature of the program that is most significant to western stock ranch management is the impetus that has been given, through this program, to the development of local cooperative grazing associations among ranch operators in areas of intermingled public and private lands. Primarily, the function of the cooperative association is the group tenure and area management of a district of lands that have a diverse and complex pattern of small tract ownerships.

In the land-use projects that were developed under the sub-marginal land acquisition program, organization of cooperative grazing associations was encouraged, and preference was given to the associations in the leasing of the acquired lands. These associations are organized under special state enabling legislation* or under a general state cooperative act. An association is empowered to lease and own land, and subject to the articles of incorporation and the bylaws, to issue grazing permits and establish grazing regulations for the lands which it controls in the district.

Through the medium of the district lands controlled by the association, a group of small and intermediate-size ranches is able to attain many of the advantages of the large ranch in the strategy of land tenure and control. Such an organization can also develop management plans for an area of rangeland that would otherwise be open and unmanaged land, or would be managed less effective-

* See Mont H. Saunderson and N. W. Monte, *Grazing Districts in Montana: Their Purposes and Organization Procedure,* Montana Agricultural Experiment Station Bulletin No. 326, 1936.

ly through individual ranch tenure than is possible on an area basis. Thus, the cooperative grazing association can be an effective private instrument for the group management of lands and intermingled ownership, including state and county lands and scattered federal ownerships.

Federal Administrative Organization. Administration of the federal lands acquired in the repurchase project areas is now handled by the Soil Conservation Service of the U.S. Department of Agriculture. Where these project lands are within the boundaries of the grazing district of a state cooperative grazing association (generally the situation in Montana, Wyoming, and the Dakotas), the lands are leased on a term basis to the association. The cooperative association establishes the regulations and issues permits to the members of the association for the use of these lands, subject to certain provisions in the lease contract with the Soil Conservation Service to insure the policies of that service in the management of the lands.

Administration of a considerable part of the federal grazing district lands of eastern Montana is now handled on a similar basis; that is, through lease and management agreement with the cooperative state grazing association. Thus the cooperative grazing association becomes the administrative vehicle for effecting organized management of grazing lands of intermingled ownership. A cooperative association can perform this function particularly well where such lands of diverse ownership are in sizable acreages and the season of use and the development and improvement of such lands can consequently be handled more effectively on an area basis than within the land holdings of the individual ranches.

GENERAL LAND OFFICE LANDS

How Leases Are Handled. Section 15 of the Taylor Grazing Act provides for the leasing of public domain lands that are not so situated as to justify their inclusion within the boundaries of a grazing district organized under the act. The act further provides that preference shall be given in the leasing of such lands to the owners and occupants of the contiguous lands to the extent necessary to permit the proper use of contiguous lands. This leasing of public domain lands that are not now within the boundaries of federal grazing districts is handled, as provided in section 15 of the Taylor Act, by the General Land Office of the U.S. Department of the Interior. (The Government Land Office is now a part of the Bureau of Land Management.)

Although the use of the lands that are handled through lease by the Bureau of Land Management is important in some localities and may be very important to certain individual ranches, such use is only 6 percent of the total animal months of grazing on federal public lands. These lands are not leased on a competitive-bid basis. The field examination offices inspect the land and determine to whom the lease should be awarded. The leases are then handled through the local United States district land office. All the western states have one or more such offices. The leases may be for a term of not to exceed ten years. While these lands are leased on an acreage rather than an animal month basis, their cost is generally somewhere near comparable with the present charges on the grazing districts — that is, $.08 per animal month for cattle and $.01¾ per animal month for sheep.

Location of the Lands. Most of the section 15 leases are in Wyoming, Colorado, California, New Mexico, and Washington. There are scattered tracts of this land throughout all the other eleven western states where there is not sufficient public domain land to warrant the organization of a grazing district under the provisions of the Taylor Act. Altogether some 11 million acres of public domain land in the eleven western states are now handled under section 15 of the Taylor Act.

POSSIBLE CHANGES IN THE USE OF THE PUBLIC LANDS FOR GRAZING

Stock ranch operators who use public land naturally believe that such use is a permanent part of a stable operating unit for the ranches. The legislation and the programs that govern the use of these lands for grazing are in varying degree somewhat contrary to this viewpoint.

Possible Future Trends in National Forest Use. Problems of plant cover deterioration and soil erosion that prevail on some parts of the western national forests make it appear inevitable that considerable acreages of these lands will have to be closed to grazing as the only practical means of protecting valuable watershed lands. There is also the probability of a growing demand for the use of the national forests for recreation and as a wildlife habitat, particularly for big game. Probably the needs of resource protection and other uses can be made compatible with the present grazing use on most of the national forest lands as a result of the management program that is now in effect.

Viewing this situation realistically, however, we see that the question of relative use values cannot be decided entirely on the basis of the local economy adjacent to the national forests. This is one of the reasons why the U.S. Forest Service has steadfastly adhered to the policy that a grazing permit on the national forests is a privilege granted to a ranch operator for a term of years and is not in any way to be construed as a property right that accrues to ranch ownership.

This probable future loss of upland range on the national forests can in some localities be offset by the development and intensive management of irrigated pastures by the ranches. There are some sites within the national forests where intensive management can increase the grazing capacity many fold to offset the closure of erosive slopes and of the concentrated grazing along upland water courses. An illustration of this development of nation-

Irrigation of favorable sites on mountain summer range, such as this one in the White River National Forest of Colorado, can sometimes offset closing steep slopes to grazing. (U.S. Forest Service Photo)

al forest grazing is the irrigation of native grazing lands on the favorable sites of the White River National Forest of Colorado (see the accompanying illustration). This type of development may require considerable investment for fencing, irrigation, and seeding. The cost of the development work should, consequently, be compared with the probable value of the increased livestock production.

Possible Changes in Grazing District Management. The condition of the ranges on the grazing district lands was downward for many years before they were placed under administration, and in the general picture this down trend has not yet been very much changed. If the trend in the condition of these ranges is to be reversed, they must receive far more management than they are now receiving. The only practical management measure for many of these lands is a reduction of present use, either fewer livestock or a shorter season of use. In some localities the reduction of use of the deteriorated lands can be offset by more intensive management of the better sites through reseeding, rotation management, and better adapted seasons of use.

These management measures will require an investment in range development, probably associated with a considerably higher grazing charge than that now in effect, and will involve some changes in the operating methods of the ranches. An illustration of this is the necessary change in the present practice of using the desert winter sheep ranges of western Utah well into the period of spring growth of these ranges (see Graph B of Figure 4).

Such a change probably will require a combination of some reduction in sheep numbers below that which has prevailed in the past, shortening the period of use by two weeks or more in the spring, and, where feasible, initiating rotation management of the range to avoid use in alternate years during the period of spring growth. Shortening the period of use of the winter ranges would require an earlier movement in the spring to the spring-fall range and the use of more supplements before and during lambing while on the spring-fall ranges. This would be the case until management measures can increase the capacity of the spring-fall ranges. Generally, the native bunchgrasses of these sagebrush-grass ranges that constitute the spring-fall sheep range will respond to management rather slowly, but practical and economic methods are being developed for reseeding the favorable soil and moisture sites of this range type.

Reseeding of the better sites in the sagebrush-grass range type with adapted introduced grasses could eventually restore much of it as a season-long cattle range, offsetting the loss of upland cattle range on the national forests of the intermountain region. Recent experimental trials show that reseeding and management can, on the better sites of this range type, eliminate the cheatgrass, increase the grazing capacity by as much as tenfold, and extend the grazing period into the summer months.*

Some Possible Policy Changes. We see, then, that there are some issues of conservation and kinds of use in the management of the western public lands. Changes in policy growing out of these issues may alter considerably the present uses of these lands for the grazing of domestic livestock. National forest grazing has been reduced, for soil and watershed conservation purposes, 25 percent in the past twelve years and 50 percent in the past thirty years. Grazing use of the public domain grazing district lands may require considerable reduction and seasonal change, for conservation reasons. These lands, and the Navajo Indian Reservation, now yield most of the sediment going into the large and costly western reservoir storages. The public domain administration has not yet made many management changes for conservation purposes.

Conservation problems are, in general, more pressing for the public lands than for the privately owned rangelands of the West. This is true because the public lands are less productive, more difficult to manage, and have had more abuse. For the stock ranch that uses public land grazing as an important feature of its land use, the hope for the future is a good cooperative working relationship between the management of the ranch and the federal agency that manages the public land used by the ranch. The recent controversies on public land grazing and conservation policies have caused a lessening of such cooperation.

* See "Sagebrush Saga," mimeographed publication of the Intermountain Forest and Range Experiment Station, Forest Service, U.S. Department of Agriculture, January 1945.

Common and Botanical Names of Range Plants

Because local usage of common names for range plants may differ, the following botanical names are given to show the use of common names by the author.

Common Name	*Botanical Name*
Alfilaria	Erodium cicutarium
Arizona fescue	Festuca arizonica
Aspen	Populus tremuloides
Big sage	Artemisia tridentata
Bitterbrush	Purshia tridentata
Black sage	Artemisia nova
Bluebunch wheatgrass	Agropyron spicatum
Blue grama	Bouteloua gracilis
Bud sage	Artemisia spinescens
Buffalograss	Buchloë dactyloides
Bulbous bluegrass	Poa bulbosa
Catclaw	Acacia greggii
Cheatgrass	Bromus tectorum
Coffee brush	Simmondsia chinensis
Crested wheatgrass	Agropyron cristatum
Curly mesquite grass	Hilaria belangeri
Galleta grass	Hilaria jamesii
Idaho fescue	Festuca idahoensis
Indian ricegrass	Oryzopsis hymenoides
Indian wheat	Plantago argyrea
Mahogany brush	Cercocarpus montanus
Mesquite brush	Prosopis (spp.)
Needle-and-thread grass	Stipa comata
Oakbrush	Quercus gambeli
Prairie June grass	Koeleria cristata
Prairie sand reed grass	Calamovilfa longifolia
Red three-awn grass	Aristida longiseta
Saltbush	Atriplex canescens
Sandberg bluegrass	Poa secunda
Serviceberry	Amelanchier alnifolia
Shadscale	Atriplex confertifolia
Slender wheatgrass	Agropyron pauciflorum
Smooth bromegrass	Bromus inermis
Snowberry	Symphoricarpos
Tobosa grass	Hilaria mutica
Western wheatgrass	Agropyron smithi
Winterfat	Eurotia lanata

Bibliography

Chapter I

BROADBENT, D. A., G. T. BLANCH, and W. P. THOMAS. *An Economic Study of Sheep Production in Southwestern Utah.* Utah Agricultural Experiment Station Bulletin No. 325. 1946.

BURDICK, R. T., M. REINHOLT, and G. S. KLEMMEDSON. *Cattle Ranching in the Mountains of Colorado.* Colorado Agricultural Experiment Station Bulletin No. 342. 1928.

CARPENTER, G. A., MARION CLAWSON, and C. E. FLEMING. *Ranch Organization and Operation in Northeastern Nevada.* Nevada Agricultural Experiment Station Bulletin No. 156. 1941.

FENNEMAN, NEVIN M. *Physiography of Western United States.* New York: McGraw-Hill Book Co., 1931. 534 pp.

GUILBERT, H. R., and L. H. ROCHFORD. *Beef Production in California.* California Agricultural Extension Service Circular No. 115. 1940.

HOCHMUTH, H. R., and MARION CLAWSON. *Sheep Migration in the Intermountain Region.* U.S. Department of Agriculuture Circular No. 621. 1942.

PINGREY, H. B. *Combination Ranching in Southeastern New Mexico.* New Mexico Agricultural Experiment Station Bulletin No. 332. 1946.

SAUNDERSON, MONT H. *The Economics of Range Sheep Production in Montana.* Montana Agricultural Experiment Station Bulletin No. 302. 1935.

————. *Cattle Ranching in Montana.* Montana Agricultural Experiment Station Bulletin No. 341. 1937.

————. "A Method for the Valuation of Livestock Ranch Properties and Grazing Lands." Montana Agricultural Experiment Station Mimeographed Circular No. 6. 1938.

————. "Western Stock Ranch Earnings and Values." *American Cattle Producer,* Denver, Colo., vol. 27, no. 7 (December 1945).

————. "Ranch Country." *American Cattle Producer,* Denver, Colo., vol. 28, no. 2 (July 1946).

SHANTZ, H. L. "The Natural Vegetation of the Great Plains Region." *Annals of the Association of American Geographers,* vol. 13 (1923), pp. 81–107.

U.S. DEPARTMENT OF AGRICULTURE. *Atlas of American Agriculture.* Washington, D.C.: Government Printing Office, 1936.

VASS, A. H., and H. PEARSON. *Cattle Production on Wyoming's Mountain Valley Ranches.* Wyoming Agricultural Experiment Station Bulletin No. 197. 1933.

Chapter II

HUIDEKOPER, WALLIS. *Modern Beef Cattle Breeding and Ranching Methods.* Printed circular. Helena, Mont.: Montana Stockgrowers' Association, 1940.

241

242   WESTERN STOCK RANCHING

KNAPP, B., A. L. BAKER, J. R. QUESENBERRY, and R. T. CLARK. *Growth and Production Factors in Range Cattle.* Montana Agricultural Experiment Station Bulletin No. 400. 1942.

Record Stockman. 1948 Annual Edition. 59th yr., no. 1. Denver, Colo.: Record Stockman, 1948.

Western Farm Life. 1948 Annual Herdsman Edition. Vol. 50, no. 1. Denver, Colo.: Farm Life Publishing Co., 1948.

Chapter III

ALLRED, B. W. *Range Conservation Practices for the Great Plains.* U.S. Department of Agriculture Miscellaneous Publication No. 410. 1940.

SAMPSON, A. W. *Range and Pasture Management.* New York: John Wiley and Sons, 1923. 412 pp.

SAUNDERSON, MONT H. "Some Economic Aspects of Western Range Land Conservation." *Proceedings of the Montana Academy of Science.* 1941.

SHANTZ, H. L. *Fire as a Tool in Management of Brush Ranges.* Bulletin of the California State Department of Natural Resources, Division of Forestry. Sacramento, Calif., 1947.

SHORT, L. R. *Reseeding to Increase the Yields of Montana Range Lands.* U.S. Department of Agriculture Farmers' Bulletin No. 1924. 1943.

STEWART, GEORGE, and R. H. WALKER. *Reseeding Range Lands of the Intermountain Region.* U.S. Department of Agriculture Farmers' Bulletin No. 1823. 1939.

STODDARD, L. A., and A. D. SMITH. *Range Management.* New York: McGraw-Hill Book Co., 1943. 547 pp.

U.S. DEPARTMENT OF AGRICULTURE. *A Range Plant Handbook.* Prepared by the U.S. Forest Service. Washington, D.C.: Government Printing Office, 1937.

_____. *Market Your Range Cattle in Best Condition.* Bulletin AWI-55. 1943.

_____. and U.S. DEPARTMENT OF THE INTERIOR. *Range and Livestock Production Practices in the Southwest.* Prepared in cooperation with the New Mexico College of Agriculture and Mechanic Arts and the University of Arizona. U.S. Department of Agriculture Miscellaneous Publication No. 529. 1943.

Chapter IV

JOSEPH, W. E. *Winter Feeding of Breeding Ewes.* Montana Agricultural Experiment Station Bulletin No. 164. 1924.

LANTOW, J. L. *Supplemental Feeding of Range Cattle.* New Mexico Agricultural Experiment Station Bulletin No. 185. 1930.

MORRISON, F. B. *Feeds and Feeding.* Ithaca, N.Y.: Morrison Publishing Co., 1939. 1050 pp.

STEWART, GEORGE, and IRA CLARK. "Effect of Prolonged Spring Grazing on the Yield and Quality of Forage from Wild Hay Meadows." *Journal of the Society of Agronomy,* vol. 36, no. 3 (March 1944).

Chapter V

KELLER, H. B. "Why Core Test?" *National Woolgrower,* vol. 38, no. 4 (April 1948).

MANN, L. B. *Cooperative Marketing of Range Livestock.* Economic and Credit Research Division, Farm Credit Administration, Bulletin No. 7. 1936.

SAUNDERSON, MONT H. "Cattle Production and Marketing in the West." *American Cattle Producer,* vol. 23, no. 10 (March 1942).

U.S. DEPARTMENT OF AGRICULTURE. *Market Classes and Grades of Livestock.* Department Bulletin No. 1360. 1942.

VAUGHN, H. W. *Types and Market Classes of Livestock.* Columbus, Ohio: College Book Co., 1942. 608 pp.

Chapter VI

U.S. DEPARTMENT OF AGRICULTURE, FARM CREDIT ADMINISTRATION, ECONOMIC AND CREDIT RESEARCH DIVISION. *Income and Expense of Sheep Ranches in Twelve Western States, 1940, 1941, and 1942.* 1943.

Chapter VII

CLAWSON, MARION. "Effect of Changing Prices upon Income to Land as Illustrated by Data from Montana, 1910 to 1936." Mimeographed circular of the U.S. Department of Agriculture, Bureau of Agricultural Economics. Issued February 1939.
GABBARD, L. P., C. H. RONNEN, and J. W. TATE. *Planning the Ranch for Greater Profit, a Study of Physical and Economic Factors Affecting Organization and Management of Ranches in the Edwards Plateau Grazing Area.* Texas Agricultural Experiment Station Bulletin No. 413. 1930.
SAUNDERSON, MONT H. "Land Values and Land Costs for Range Livestock Production." *National Woolgrower,* vol. 28, no. 12 (December 1938).

Chapter VIII

FLEMING, C. E., and C. A. BRENNEN. *Ranch and Range Balance. The Public Lands and Ranch Stability in Nevada.* Nevada Agricultural Experiment Station Bulletin No. 142. 1936.
ROTH, ARTHUR. "A Graphic Summary of Grazing in the Intermountain Region." ("Seasonal Relationships of Lands Used in the Production of Range Livestock," Part II.) Mimeographed circular of the U.S. Department of Agriculture, Forest Service, Intermountain Forest and Range Experiment Station, Ogden, Utah. 1941.
————. "A Graphic Summary of Grazing in the Intermountain Region." ("Present Use of Public Land Grazing Resources by Size and Type of Livestock Enterprise." Part III.) Mimeographed circular of the U.S. Department of Agriculture, Forest Service, Intermountain Forest and Range Experiment Station, Ogden, Utah. 1942.
SAUNDERSON, MONT H. "Some Economic Aspects of the Upland Watershed Lands of the Western United States." *Journal of Land and Public Utility Economics,* vol. 15 (November 1939), pp. 480–82.
————. "The National Forest Officer Looks at Resource Values." *Journal of Forestry,* vol. 45, no. 4 (April 1947).
————. and N. W. MONTE. *Grazing Districts in Montana: Their Purpose and Organization Procedure.* Montana Agricultural Experiment Station Bulletin No. 326. 1936.
U.S. DEPARTMENT OF AGRICULTURE, FOREST SERVICE. "U.S. Forest Service Manual for the Protection and Management of the National Forests," Range and Wildlife sections. (U.S. Department of Agriculture Forest Service Manual, Vol. III.)
———— INTERMOUNTAIN FOREST AND RANGE EXPERIMENT STATION. "Sagebrush Saga." Mimeographed. Issued January 1945.
U.S. DEPARTMENT OF THE INTERIOR, BUREAU OF LAND MANAGEMENT. *The Federal Range Code.*

Index